Hands-On HORTSCIENCE:

Practical Investigations & Experimental Designs in Agriculture and Biology

Carrie L. Whitcher Moniux, PhD
California State University, Chico

KENDALL/HUNT PUBLISHING COMPANY
4050 Westmark Drive Dubuque, Iowa 52002

ISBN 978-0-7575-5296-0

Printed in the United States of America
10 9 8 7 6 5 4 3 2 1

Dedication

For Claude, my husband, partner, and best friend, thank you for your love and support

Contents

Preface

This Hands-On Hortscience laboratory workbook was developed to assist students in understanding how to "do science" with actual experiments and research. It assumes no prior scientific methodology or statistical knowledge on the part of the student and uses examples in both plant and animal sciences to introduce research principles. This workbook is intended for use in a one-semester or one-quarter introductory course in agriculture or biology but may be useful for graduate students or advanced placement high school students in these areas.

Scientific research is a fluid process that has no standard "method" but consistently revolves around observation, data collection, and interpretation. The overall goal of this workbook is to give students the tools they will need to read peer-reviewed literature critically, understand how to develop organizational skills, develop original research objectives, plan and execute an experiment, and analyze the research results. While statistical methods will be used, emphasis will be on understanding how those methods help students make better, more informed decisions, not on the computations.

Each chapter has exercises where students will have an opportunity to critique peer-reviewed research and develop scientific investigations through hands-on labs. Through learning by doing and becoming an active part of the research process, it is hoped that students will develop an appreciation for the time and effort involved with producing credible scientific research results.

Acknowledgements

The author is grateful for inspiration and assistance from colleagues with whom she has shared ideas about teaching horticulture science, thank you for your suggestions: Ms. Geraldine Maxfield, Mr. Matthew Kent, Ms. Heather Wren, Dr. Patrick Doyle, Mr. Wes Schager, Dr. Richard Baldy, Dr. Nancy Carter, Mr. Tip Wilmarth, and Mr. Claude Monlux.

Special thanks go to Linda Chapman and Dr. Frank Forcier, editors, and all of the staff at Kendall/Hunt Publishing.

I am particularly grateful to the faculty members who have inspired me throughout my educational career, being the role models I truly admire. You gave me your time and considerable feedback during the review of this and previous works. Thank you for being who you are and loving what you do in the classroom and in the field. You are real teachers, and I hope to be like you someday.

Dr. Gary Briers, Texas A&M University
Dr. Tom Cothren, Texas A&M University
Dr. Frank Hons, Texas A&M University
Dr. Dan Lineberger, Texas A&M University
Mr. Claude Monlux, Chico High School
Mr. Quentin Nakagawara, Butte-Glenn Community College
Dr. Wes Patton, California State University, Chico
Mrs. Dorothy Ramon, Chico High School
Dr. David Reed, Texas A&M University
Dr. Kingsley Stern, California State University, Chico
Dr. Jayne Zajicek, Texas A&M University

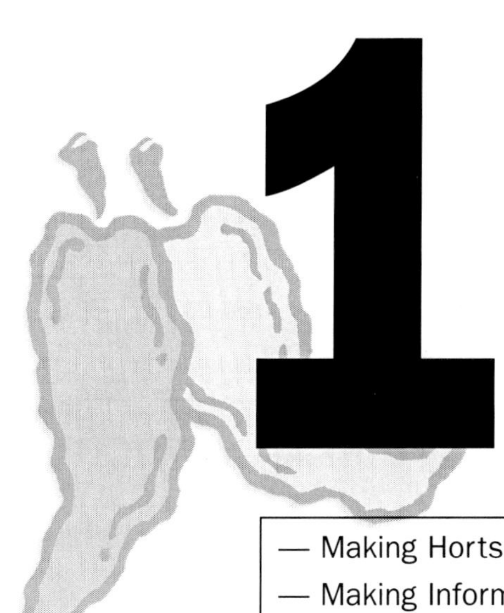

1 What is Hortscience?

- — Making Hortscience Real and Meaningful
- — Making Informed Decisions
- — Introduction to Agriscience and Hortscience Research
- — Why Textbook Learning Does Not Always Work
- — Exercise 1-1: Student Information Sheet

Why Should You Read This Book?

One of the graduation requirements of almost all students in their educational career is some life science course. It can be either a year or a single semester long and is supposed to give the student an opportunity to learn about the living world around them: plants, animals, bacteria, fungi, and other organisms. The problem with most life science courses is they rely on learning many new terms and concepts, which is not anything new to you, but usually they are not relevant or meaningful to your life. In this text, we are going to show you how hortscience and agriscience can help you make better decisions about the products you have to choose from and how to critically read and interpret the literature (or commercials and advertisements) to become a better consumer.

Making Hortscience Real and Meaningful

What we are going to do in this class is take some plant cultivars (you will sometimes see this abbreviated as cv) or varieties that you might find in your home landscape, garden, or on a farm and test some new products or management techniques on them to see how the plants perform (e.g., different fertilizer types). We are going to choose the variables we want to test, set up the experiments, collect and analyze the data, and make some decisions based on the experimental results. From these findings, you will make some recommendations about the products the plants used, and critique the experimental design. These results will be written up in a research lab report by each student and submitted to the instructor toward the end of the semester. The research lab report is like a mini-journal article, so part of this book will be devoted to helping you write the lab report in the correct manner. The remainder of the book will help guide you through the research process and answer the question, "Why do scientists do whatever it is they do?"

If agriculture (namely hortscience) is not your background or specialty area and you are taking this class to fulfill a science graduation requirement, you can still use the same techniques to help you solve problems and make decisions in other areas. If you are an agriculture or science major, this book will help you develop an experiment in agriscience, animal science, biology, botany, chemistry, or any of the physical sciences. Science is science, whether it is called hortscience, agriscience, biological science, or physical science. Learning some tips, timesavers, and technology in scientific research and writing are vital to making science work for you. Remember though, you need to do your part and connect what you are learning back to your real life. Only through this connection and hands-on activities can you learn. Your instructor can only provide the experience for you; you have to undergo the experience to learn for yourself.

If you are still wondering how this relates to your life, somewhere along the way in your professional career after graduation you will probably need to write. You may need to compose a summary of your work for your employer, create a bill of materials and budget for a project, prepare a spreadsheet of sales data, compile some project results obtained by several of your employees, or provide background information to make some important company decision. Are you prepared to handle those tasks? If you are not, that is okay. This class will help you learn how to do those tasks and many others. Even though this is a life science course, we will teach you how to combine these decision-making skills with those you are learning in your major. Remember, what you learn in this class and from this book can be applied to everything else around you.

Making Informed Decisions

Do you make important decisions on a daily basis? Of course you do. As a consumer you have literally thousands of choices to make every day: what to eat, what to wear, how to get from point A to point B, where to buy gas, what classes to take, where to find a job, and so on. A main decisive factor for most people in making choices is economics: your choice is probably based on cost. You usually select the lowest cost and that is fine. We do the same thing in hortscience; we want to grow the best quality fruits, flowers, and vegetables with the least amount of inputs. As a consumer, this should interest you. You already know about science but probably never made the connection between "doing science" and making decisions before now. In this class you will learn how to compare products to determine the best buy and calculate costs, inputs, and measure productivity. These skills will help you become a better manager, administrator, technician, business owner, and quite possibly even a more productive family member. You will learn how to communicate those results to others better, and develop a bit of "salesmanship" to promote your findings.

Introduction to Agriscience and Hortscience Research

The definition of science can be found in any common dictionary and is generally defined as a branch of knowledge or study dealing with a body of facts or truths systematically arranged and showing the operation of general laws (e.g., the biological sciences). It has also been defined as systematic knowledge of the physical or material world gained through observation and experimentation (this is my personal preference).

Too often students take a science course (life or physical) and end up memorizing a bunch of terminology, running some experiments, and answering questions about the procedures in the experiment. They give some memorized terms back on an exam, and that is it, their science requirement is complete.

Boring! Maybe that is one reason why so many students are so turned off by science; it has been presented as a bunch of random memorization that does not mean anything, and it does not relate to anyone's real life. Anyone who teaches science courses as a vocabulary class is really doing the scientific community a disservice, there is so much more to it. Science is how we learn about new phenomena and how to make better decisions. Since you are taking this class, it is obvious that you want to learn more—more about the natural world, more about making better decisions, and more about yourself.

This class will focus on research found in agriculture, particularly horticulture and plant science, but will also provide examples in the animal sciences and ag biology. I promote the use of plant research for several reasons: plants are fun to watch because they grow quickly; they are inexpensive to grow; they can grow in any environmental condition or in a greenhouse; and they can be used for food, feed, flowers, or medicine. You probably have had some experience with plants and animals through your care of houseplants, pets, gardens, or if you have an agriculture background.

Since this is a course in agriculture, most of the examples will be related to commercial agriculture for the majors who take this class and will continue on with advanced, upper division coursework. You are not an ag major? Do not worry, I expect you either are or will become a homeowner, so many examples will also be provided for the general taxpaying consumer. Most of you already cook so developing an interest and awareness in selecting fresh produce and herbs can advance your culinary skills. Growing your own garden is a sustainable practice where you will learn how to grow your own food and appreciate the time and labor invested in keeping a safe food supply. Plus, gardening and cooking are fun!

Why Textbook Learning Does Not Always Work

As mentioned before, students need to understand that science is a way of making comparisons, observing new phenomena, testing new methods and techniques, and ultimately, making unbiased decisions. Science is not just learning structures and their associated functions, it is about making connections between why things are the way they are and how they affect everything around them. In this class you will learn through many different ways: hands-on, linguistic, visual, auditory, kinesthetic or naturalistic, to name a few. This course will allow you to learn through many different avenues with the hands-on labs and activities through investigation and experimentation. You will have the opportunity to get as involved with the course as you want to! By getting "down and dirty" in each lab, you will learn since you will be doing the actual work. In the end, you will compose an authentic research lab report that will allow you to take what you have learned in lecture (structure, function, and theory) and apply it to an applied research problem that you conducted in the lab. This lab report will contain all of the regular features of a refereed journal article: title, abstract, introduction, materials and methods, results, discussion, and literature cited sections.

Are you worried that you are a bit rusty when it comes to scientific writing or have never done it before? Do not worry, that is the point of this book. I will walk you through the steps so you can gain confidence in your research and writing skills and give you tips and timesavers along the way. Science has its shortcuts too, knowing what they are can save you a lot of time and effort.

If you were asked to memorize the parts of your car and all of its components, would it help you drive around town better? Certainly not! So why do we ask our students to memorize plant, seed, and flower anatomy, and yet expect them to know why farmers and ranchers do what they do to farm commercially? How can we expect the public to know how to grow vegetables in a garden from only learning about those parts and pieces? We cannot, and we should not. Would it not be beneficial to learn how to make the connection between the data we know and why we need to do something such as deciding which fruit tree varieties grow best in our area? When you look at the data from an experiment you conducted, you will learn how to interpret those data and make decisions and recommendations from them. That is why you are in college, right? To learn how to make better decisions, get informed, and learn how to learn more? Is that what getting an education is really all about?

Someone always asks if they are going to learn the "scientific method," a term which I rarely ever use. The reason for this is that there is not always a definite method to test new ideas or make comparisons. I like to stick with the terms investigation and experimentation. These truly fit what we are going to be doing when we want to figure something out, analyze data, and make decisions about some topic or problem. It is important that as you read this book, you constantly remind yourself why you need to know this material and how it will relate to and benefit your life. Perhaps you already are/will be owning your first home and maintaining the landscaping or desire to grow a vegetable garden. Maybe you are majoring in agriculture in school and want to learn how to make better decisions about selecting a fertilizer or pesticide for one of your crops. Regardless, both situations will benefit from using the techniques found in this book and learning how to communicate your findings to others will make you a better employee, manager, administrator, or educator.

It would help your instructor to know more about your agriculture background or if you have any experience in the plant or animal sciences to better tailor this course to your abilities. Take a few moments to complete the first assignment, honestly answering each question with examples from your own life. Remember to follow the guidelines for handwritten assignments as specified in the syllabus.

2 Some Basic Scientific Research Concepts

What Are We Trying to Accomplish?

The reason for this book is to help you develop higher order thinking and communication skills in agriscience research. Specifically, you will learn how to 1) develop a specific topic or problem of interest; 2) conduct individual research in published, peer-reviewed literature; 3) apply previously published knowledge to your original research experiment; 4) analyze data and evaluate the results; and 5) communicate your research findings in a scientific lab report.

Developing a Hypothesis

Now that you have an idea what you are going to learn by conducting experiments, you need to develop your research ideas and objectives. What do you specifically want to learn about plants or animals? Nutrition? Reproduction? Management strategies? When you come up with an idea to study and why, it is time to turn it into a hypothesis. A **hypothesis** is a best guess why something is happening. You might have some ideas why something works (or does not work!) but need to test your ideas to be sure.

When it comes time to analyze the final data you have collected at the end of your experiment, there are specific statistical tests we will use. These tests do not measure your actual hypothesis (best guess or estimation); they will actually measure a related null hypothesis. The **null hypothesis** states there will be no difference found between the groups (or treatments) you are testing. An **alternative hypothesis** states a difference will be found between the groups (or treatments) you are testing.

It is a good idea to get into the habit of writing out the null hypothesis (**Ho**) and the alternative hypothesis (**Ha**) as you are planning the experiment. This will define what you are trying to test, help you determine what type of experiment to set up, and will assist you in writing your objectives for the study.

When you look at the Ho, you will see it states there is no difference between the product or procedure we want to test and what we are using now. Is this what we really want to find out? Of course not, the Ha is what we are really physically testing, to see if we can reject the null hypothesis and confirm there is a difference between the groups (treatments). Remember that we cannot prove the alternative hypothesis, only support it. For example:

> Ho: Potting soil type has no effect on geranium height.
> Ha: Potting soil type does affect geranium height.

Here are some other examples:

> Ho: Fertilizer type has no effect on number of flowers grown on geraniums.
> Ha: Fertilizer type has an effect on number of flowers grown on geraniums.

In livestock, an example could be something like this:

> Ho: Milk replacer brand has no effect on dairy calf weaning weight.
> Ha: Milk replacer brand does affect dairy calf weaning weight.

It may be confusing to have a null hypothesis (which sounds negative) but we need to be reminded that we never "prove" anything in science, just generate more data to support or refute current theories. A theory is a statement that has been supported with examples so often that we accept it as fact.

If our data support the null hypothesis, we might see something like this, "There is no difference on weaning weights between milk replacer formulas in dairy calves." That is all we can say and hope to add to the scientific knowledge base. If our data end up supporting the alternative hypothesis, it would sound something like this: "Our data show there is a difference in geranium heights grown in different potting soils, therefore we reject the null hypothesis" or "These data support earlier research that there is a difference in number of flowers grown on geraniums by type of fertilizer applied."

Try your best to never include the word "prove" in your lab report or scientific papers. We only support or refute our hypotheses or existing theories, never prove. This will be discussed again when we get to data analyses later on in the book.

Developing a Specific Topic and Objectives

What is it specifically that you want to learn more about? Do you want to test a new product you have read about? Are you making a choice between two or more products? Do you want to compare a new procedure against an old reliable standby? Take a minute and write down specifically what it is that you trying to accomplish and write it out so that it is measurable. This will become one of your **objectives**. Creating several objectives in your study is probably the most important thing you are going to do as you design your experiment. It narrows your focus, controls the size of your research, and most importantly, points you in one direction.

An example of a poor objective would sound something like: The objective of this research is to study plant growth using different fertilizers. Why is this statement a poor objective? It does not give the reader any idea what is going to done or what is going to be tested (it sounds like fertilizers, I suppose, but we do not know what kind, brand, or amount). We know it will involve plants but what kind of plants, flowers? Vegetables? Fruit trees? What is going to be measured in this experiment, will it be height or weight? Flower number? Fruit yield?

No, this objective does not specifically address anything and if fact says its purpose is to study. Does this mean all that we are going to do is study background information on plant growth and fertilizers and not actually grow anything? This objective does not tell us much and is confusing to the reader, making it a poor example.

A better objective that is more specific and meaningful would be something like this: The objective of this research is to determine the effects of fertilizer rates on plant height in sunflowers. This sounds better because we are actually going to determine something (plant height) based on different fertilizer rates on sunflowers. It is better, but it could be more descriptive still. What kind of fertilizer? What kind of sunflowers? Where are the sunflowers going to be grown, outside or in a greenhouse? Being more specific helps us create and tailor our experimental procedures to what it is exactly we want to learn.

Let us rewrite the objective again even more specifically. How about this: The objective of this research is to determine the effects of five Miracle-Gro® fertilizer rates on plant height in sunflower 'Big Smile' grown in potting soil. Sounds much better, does it not? Yes, it does sound better, but it could be written still more specifically. By the way, are you familiar with the terms effects and affects? They are often confused in scientific writing. An easy way to distinguish the two when you are reading and preparing your research is this: effects are like results, and both have a letter "e" in them. Affects are like changes, they both have a letter "a" in them. Our objective is to determine the results of fertilizer rates (not the changes of fertilizer rates) on sunflowers 'Big Smile.' As you read journal article titles during your literature review, take note of the use of these terms and how the authors use them appropriately.

Getting your topic narrowed down into objectives is usually the hardest part of any scientific research, but once you have an idea that you want to pursue, the rest of the set-up should be relatively easy. A word of caution though, all too often we get so enthusiastic about our research that we start missing the forest for the trees and undertake too much. When that happens, back up, look at your objectives, and go from there. Constantly ask yourself throughout your research if you are biting off more than you can chew. Even graduate students have to remember this. Do your research in chunks, manageable projects, so you do not become overwhelmed and get burned out.

What Is Known About the Topic? Conducting Background Research

Once you have the topic of interest narrowly focused for your research study, it is time to see what is already known about your objectives and hypothesis. Maybe someone has already determined what it is you are trying to learn, so why reinvent the wheel? Use their results and make your decision and move on.

It is possible there is information out there that is relevant to your study but is distantly related. That is fine, use what someone else has learned to your advantage. You could adjust your research procedures using already known methods. Conducting a thorough literature review (looking for background information on your topic) serves many purposes:

1. You are trying to see what is out there and to verify that your work has not already been done by someone else,
2. You can see what experts in your field are saying about your topic and if there is agreement,
3. This is an excellent way to gather references from articles for your lab report,
4. You will gain an understanding of what is involved with executing the research by reading other authors' materials and methods sections, and

5 There are probably model journal articles that will serve as a guide to help you write your final lab report. The latter is important when it comes to formatting your actual paper.

Locating background information about your topic (called a literature review) for your lab report can be accomplished using many resources including articles in peer-reviewed scientific journals, professional industry magazines or websites, books, and commercial handouts and promotional materials. Your task during the **literature review** is to learn what is currently being used in your specific industry or field of interest. Looking through professional magazines or talking to industry professionals can lead you to new or upcoming products or procedures. Have these products been used on your specific plant or animal species? Can you substitute or alter the amount of a new product or an established one that is already on the market? Find something for your research that will keep your interest and is relevant to your life. If you do so you will be successful. If not, you will probably fight it and will most likely fail in completing the project. Do not fall into that trap!

Students always ask about using search browsers on the World Wide Web (WWW) and online references. Here is a tip and word of caution: online references are only as good as their credibility. You will need to verify if WWW information is credible. It must meet several of the following criteria to be a credible source: is it objective, unbiased, current, and refereed? Do your homework and verify that your research references meet these standards. If they do not, that is alright, you can use these references for inspiration and foster ideas, but you should not use them as background research references. Most instructors (this one included) will not allow encyclopedia-type website references in a college lab report or scientific paper. The problem with using WWW references (and I do not mean refereed journal articles downloaded electronically) is that they are usually not credible sources of information. A library tour usually includes tutorials on researching for credible resources and periodicals. Take advantage of the librarian's knowledge!

Credibility and verification are how science separates itself from pseudoscience: information and data taken from anonymous websites, surveys, and polls have no way of being verified and therefore are not credible. Anyone can go into an anonymous website and take the survey or add to the website; therefore there it is not a credible source. Peer-reviewed work is credible because it allows the work to be critically "judged" by experts in the field.

How Do You Learn to "Do Science"?

Learning how to "do science" by conducting background research is like learning how to ride a bicycle. Remember when you learned how to ride your first bike? You watched others do it, you studied the bike itself, you probably sat on it, practiced your balance, and started pedaling for small distances. Through trial and error you probably fell off a time or two but eventually you gained balance and began riding on your own. Doing background research and designing an experiment are no different than learning how to do something for the first time. Where do you start? You start by watching what others are doing, seeing how they organize their plans, and how they executed their experiments. A good piece of advice is to start at the end...of another experiment. Where do you find such information? That is easy; peer-reviewed/refereed journal articles have all the information you need to get started.

Consider what a refereed journal article is: it is a peer-reviewed and scrutinized summary of one person's or a group's experiment, results, and interpretation. Journals are the backbone of communicating scientific research and provide a commitment to seeking the truth about some phenomenon or observation. In developing an idea to study, someone saw something or read something and got an idea to study it further. He or she developed the experiment to reduce bias and increase objectivity, collected and analyzed the data, and presented the results; comparing those results back to what was already (or not already) known. A panel of article reviewers (usually select university faculty), read the articles and judged (graded) them, and decided if the work was worthy of publication. Several anonymous reviewers read each article to provide appropriate feedback to the author. If the work is original and contributes new information that is appropriate to that journal, the article may be published. Authors are prepared to correct versions of their articles and may have to resubmit them to other journals that may be more appropriate for the topic, if that is a problem. All in all, the work is typically reviewed by experts in the field in a blind evaluation to provide feedback. No one gets to publish his or her opinion without presenting the background and the facts, and this is what sets scientific literature apart from everyday media, and this is why we should not use non-refereed WWW sources in scientific research. They lack credibility and objectivity. Anyone can put information up on the web. Anyone can claim their product is 99.9% effective. But are they accurate and reliable? Were they conducted with precision? Were most sources of error controlled and reduced? Probably not, and this is why they cannot be trusted, they lack objectivity. The peer-review process helps protect objectivity and credibility in scientific research and dissemination of results. It is not a perfect system but does its best to protect authors and researchers and their intellectual property (their research).

Communicating Your Research Findings

What effort goes into writing a scientific lab report or peer-reviewed journal article of high quality? You will know since you will put the same effort into writing your scientific lab report for this class. The only difference between them is that a journal article is based on a project of a much larger scale.

Like a journal article, your lab report will be the result of many hours of hard work and constant revision. As you look at several articles from different journals you will see that the format of each one is generally similar with differences in font, layout, citation style, and punctuation. Make sure you follow the directions given by your class instructors when submitting written work. They may specify that you follow the format of a specific journal. Follow those directions to the letter, and you can usually figure those out from reading articles from that journal or style manual. Take the journals *HortScience* and *Journal of Animal Science*, both are quality, peer-reviewed journals with wide varieties of topics covering many species of plants and breeds of livestock. Their formats are somewhat similar and their literature cited sections are almost identical. The thing to remember is that almost all journals require the same information, they just put into their own style and format.

You are going to read several articles from different journals that cover a specific topic. This will be a common theme in your college courses. Many professors will assign readings from current journal articles in their classes in most majors at the university level. The advantage of reading journal articles over textbooks is you get current information in a convenient format (e.g., small readings, PDF articles, electronic copies). In this day and age, knowing how to access electronic documents is an important skill to have, and it will separate you from your competition when you apply for employment upon graduation.

As you read through several articles, you might not know some of the terminology or technical background. Do not worry about that right now. Take some time to compare the articles and determine what is obviously different and stands out to you. Is it the format of the title? The way the author(s) names are listed? Is there an abstract or summary provided at the top of the article? What are the differences and similarities of format, font, and style?

All of these elements and many others are very common to journal articles but what connects them is the foundation of the peer-review process to promote academic freedom. With this process, authors make every effort to ensure their work is protected from being stolen by those who did not do the work. Being able to share one's work and receiving full credit for doing so is another hallmark of the scientific community. Make sure you always cite your sources when quoting or using others' research. **Plagiarism** is the worst of academic sins, where one takes another's words or ideas and presents them as their own. We expect you to find outside references and cite your source, that is the point of research! Being able to trust colleagues to not steal one's research ideas and results is critical to the exchange of ideas in the scientific community. Plagiarism will not be tolerated in this class and all violations will be reported to the office of student judicial affairs. Do not risk getting expelled from the university because you did not do your own work, it is cheating. Realize you are here to learn how to learn, and that means telling us (your instructors) where and how you found your sources. Here is an interesting story that happened many years ago. I had a student who did not put her sources in her research paper, she wrote a lot of good ideas and had a very strong thesis statement but nothing was cited to back up her ideas. When I questioned her about it (I could not assign full credit to such a paper) she told me she did not want me to think she was dumb for "not knowing anything by having so many references" in her paper. Wow! I was stunned. I took the time to explain to her that knowing where to find information, knowing how to share that information, and making connections between what is known and what is not is the point of writing research papers. We instructors expect you to have many references and citations, that's why you are in college! Needless to say, her next paper was much better and she told me that she found it liberating to know she was not "dumb" for not citing her sources.

The following exercise will assist you in looking through several journal articles to see what they have in common and teach you to look at specific parts of a scientific paper. This will familiarize you with the required parts of your lab report, which will follow the format of a standard journal article, and will include all of the sections that are generally found: title, abstract, introduction, materials and methods, results, discussion, and literature cited.

Brainstorming: Putting Your Research Ideas Down on Paper

When developing new ideas for research, it is helpful to keep track of your notes and ideas in a bound composition notebook. These notebooks have the advantage of being inexpensive, hardbound, college-ruled, and easily portable. All too often some important reference or data set has been misplaced or lost when a page was torn out of a loose-page binder or spiral-bound notebook. Make it a rule to never remove pages from your composition lab notebook and keep it for research notes only. Write your name and contact information in it in case it gets misplaced, then it can be returned to you.

An easy way of gathering references and potential sources of information/research ideas is to review other articles' literature citations pages. Of course you will need to read current articles to see if there are any new developments in your area of interest but do not reinvent the wheel if you do not have to. Take a look at what other researchers have done in your area of study and take note if the protocols in the materials and methods sections can be modified to fit your experiment. You should be able to come up with a plan of action by seeing what has already been published, and in particular, what has not. Be aware that you may need to create your own procedures for your specific research problem so the time you invest in reviewing published work may assist you in creating these new procedures. Remember, do not reinvent the wheel!

Are you familiar with each of the sections generally found in journal articles? As mentioned before, they generally include the following items: title, abstract, introduction, materials and methods, results, discussion, and literature cited.

If you are new to reading peer-reviewed journal articles in any discipline, here is a simple way to review several journals if you are lacking time and need to verify if several articles can contribute to your topic of interest. First of all, thoroughly read the title; is this something that may be related directly or indirectly to your topic? Then do the following for each article, taking notes and sorting the papers into three categories, a "keep" pile, a "possibly save" pile, and an "info not relevant" pile:

1. Read the Abstract. It is a summary of the article's results. It paints the big scientific picture for the reader, relates the work in this study to this big picture, and states how this work may have changed the understanding of knowledge in this field.
2. Read the last paragraph of the Introduction. It usually contains the objectives and reasons for creating the study. It is very specific and points out what the goals of the research are to the reader. Do these objectives relate to your work?
3. Read the first sentence of each paragraph in the Discussion. These sentences highlight the main points that came from this study and whether the objectives were met. You can also find other relevant information here toward the end of the paper; perhaps something new occurred that the authors did not expect or recommendations for future work. This is very important if you are conducting a similar research study. Can you take this author's suggestions and turn them into something you can do?

As you start your literature review to locate information for your lab report, you will typically start reading the article title which is followed by an abstract (or summary), key words, and author(s)' names. This is followed by an **Introduction** which gives background information from several expert authors in the field. The introduction presents a subject in three to five paragraphs beginning with some general biological information and ending with some specific objectives of the experiment. These are the same types of objectives you will have for your research and are related to the background information from the experts cited in the introduction.

Take note: what do these experts contribute to the background in the introduction? Is there agreement among these experts? How do they differ in opinion? What will your research contribute to this area of knowledge? These questions are vital to writing the introduction and discussion sections of your lab report. It is a good idea to keep track of your sources with a computer spreadsheet or database program. There are commercial computer programs available (e.g., EndNote®) as well as shareware and freeware on the WWW that can help you organize your references and build citations and bibliography pages in your report. This software is used as a database to organize and keep track of your citations. A benefit of this type of software is that is can actually insert a reference into your paper (where you tell it to do so) and builds your literature citations page as you write your paper. Yes, that means you do not have to actually write your literature citations page, it just appears as you insert citations into your paper, fantastic!

The **Materials** and **Methods** section of an article (and your lab report) is typically found next. It provides details of how the study was organized, its location, and how it was executed. Here you might find important references for purchasing supplies, materials, and equipment that you might need for your study. You will notice that this is not a list. It is a written description of the set-up and implementation of the research. If you need to read papers quickly during a literature review, it is probably best to leave this section for last, unless you are looking for specific procedural advice or suggestions for supplies and materials. Some materials and methods sections require that the city and state of manufacture of a product be given. Take note of this so you can keep track of your products used so you can include this information in your report.

The **Results** section is generally found next in a scientific paper and may be combined with the discussion section in many journals. Sometimes it comes right after the introduction section, since it is an important part of the paper. The results section gives just that: the numerical or descriptive results found from the research. You will probably find tables, figures (graphs and pictures), and statistical summaries in this section. Usually the data are given in the text of the results description but the authors refer to specific data in tables and figures (using parentheses) that they want the reader to specifically notice. There is a fine line between stating facts and drawing conclusions from your results. Save conclusions for the discussion section. If you have a combined results/discussion section, you have more freedom in this area. Consult your model journal article that your class is following for clarification. For your lab report in this class, you will include tables and figures (graphs) that will highlight your significant results to your reader. Are you familiar with creating tables and graphs? If not, more details on how to create these with a computer spreadsheet program (i.e., Excel) will follow later on in this book.

The **Discussion** section of any paper is probably the most important part of a scientific journal article and your lab report because the conclusions from the data and implications are given to the reader in this section. You will see that the author discusses her/his results, makes conclusions, and connects those conclusions back to the original objectives found in the introduction. The author also attempts to support or refute the background information made by the experts in the introduction as well. As you read the discussion, ask yourself whether this new research changed opinions or gave speculations about currently known beliefs in this area. Notice how the author presents these findings in the last sentence of the abstract.

Do your results support or refute what you found during the literature review for your introduction? You will need to consider this when you compose your lab report. This is the basis for your discussion section. Keep this in mind as you are writing the abstract.

There may be references to figures and tables in the discussion to point out highlights from the research. Sometimes the author will criticize the research and give limitations of the experiment; perhaps even recommending what should be done in future studies. This is valuable information because it can give you some ready-made ideas about some objectives you can include in your experiment. Take note though, usually if future work is recommended, it is probably already underway by that author, but you can gain some insight of seeing what the author thinks the next step should be.

The final part of the article and your research report is the **Literature Cited** section. The references cited throughout the paper are found here, generally in alphabetical order by primary author's last name. Some journals use numbers to refer to authors throughout the paper, some use last names. Be sure to follow the format of the model journal specified in your assignment, and remember, for the great majority of scientific writing, **you cannot put a reference in the literature cited section if it did not appear in the paper!** Yes, that means you cannot just include a long list of references at the end of your paper and expect the reader to know what each reference contributed to your work. It is all about credibility; and citing your sources as you compose your paper is what makes it a credible piece of literary work.

Here is a timesaver you can do when proofreading. Use your literature cited section as a check-off list to find your citations throughout the paper. As you go through the paper, check off each citation in the text and find that citation in the literature cited section. Check it off there as well, verifying the information (names, dates, page numbers, etc.) are correct. This method of double-checking verifies you have everything cited you thought you included in the paper and also those that you know you did cite!

Developing a Timeline

One of the basic criteria for creating an experiment is working with a timeline. What is your window of opportunity for this research? Do you have a deadline that must be met? Is there a specific growing season for your crop or age of animals you need to use? Are there any specific materials you need to obtain or procedures you need to learn to perform your experiment? Looking at the materials and methods sections of several papers will help you answer these and other questions including start and end dates, harvest information, and equipment and materials used. Many materials and methods sections in journals are required to give the manufacturer, city, and state of materials and equipment which you can take note of for future reference.

If you are conducting an experiment for a class and need to develop a timeline, it is best to work backwards from the due date. What is the final deadline for your lab report or final paper to be completed? Start with that date and back up at least a week to allow for word processing, editing, and proofreading. Then consider the following items in developing your timeline, adding time in days or weeks that you will need to complete each deadline. Use a calendar to determine the due dates for each item, working backwards from the final due date. This will help you get organized with your work and put the project into perspective. On a calendar, pencil in tentative due dates for the following:

1. Final deadline – this is the due date to submit the lab report or research paper to your instructor in its final form.
2. Deadline to complete the rough draft – allow yourself time to make corrections and final word processing of all parts of the paper.
3. Write the Abstract or summary – you will need the Results and Discussion sections completed to write this part of the paper.
4. Write the Discussion – this part of the paper can take a long time to prepare, consider taking at least 1–2 weeks for this section. Back up your work often!
5. Write the Results – include any tables, graphs, or other numerical data relevant to the work. Making graphs takes time so become familiar with any software you may need to use ahead of time. Entering data into a computer spreadsheet can be time-consuming. You do not want to rush during this step because mistakes can ruin your results. Make sure you back up all of your work in several places often!

6. Complete the Literature Cited section – give an alphabetical list of all references cited in the paper. You might still be adding to this section if you continue with your literature review long enough. Be sure to check off your citations in your paper against this list to verify they are all there. Consider using bibliography software.
7. Develop the Materials and Methods section, including dates, materials, equipment and procedures used, and any other protocols relevant to your study. You need to make this section complete enough that someone else can to replicate your study. Make completing this section easy for yourself by keeping track of all materials in your composition notebook as you develop your experiment. Trying to locate names, dates, and other information well after the fact can be frustrating. Keep track of everything as you go along, stapling handouts into your notebook so they are not lost.
8. Develop the Introduction – describe how your work fits into the big picture of agriculture, biology, and science. What are the experts in your field saying about this topic, do they agree on the topic or are there conflicting opinions in the literature? What is known (or not known) and how is this relevant to your objectives and what you hope to accomplish? Know what you know and know what you do not know; meaning, tell the reader what has and has not been learned about the subject yet. Provide measurable objectives, ones that you can truly achieve with your research and abilities.
9. Create a Title – one that is descriptive enough to capture the readers' interest and fully describe what you have found. Sometimes the title gets changed after writing the Results and Discussion sections in the paper, and that is alright.
10. Develop your objectives – do your homework and conduct a thorough literature review. What topic sounds interesting and relevant to your life? If you buy into the project and make it your very own, you will want to work on it and will be successful. If you do not, you will fight it and will probably fail to complete it. This is common with all researchers and that is why it is okay to switch topics if you have the freedom to do so. In this class, topics have already been chosen by the instructor.
11. Read the research topic assignment – make a list of the requirements and final due date. What is the format and where/how does it need to be submitted? Is there anything special you need to know (e.g., does it need to go through Turnitin.com for review) prior to submission? Is it a group paper or an individual's paper? In this class, we will share the results but everyone must submit her or his own paper.

Always allow yourself more time than you think you will need. It is better to overestimate the amount time necessary than to be caught short!

Learning to be patient is probably the hardest part of learning about experimental research, since nothing happens overnight. Be realistic in your expectations of your goals and abilities and always double-check with your instructor or supervisor before starting any research experiment. Your supervisor may know shortcuts and have resources for you, take advantage of her/his experience and knowledge! Other students and instructors will have ideas too; consult anyone who might be able to give advice about your project. Remember though, advice is free; you do not always have to take it. It is your project.

Remember to allow time for growing plugs from seed, receiving plants or livestock from sales brokers, and growing cultures in the lab. Are you ordering in specific materials and supplies? Be prepared to wait for emails from sales representatives and returning phone calls, to allow paperwork for purchase orders and payments to be processed by accounting staff, these can take up to several days so plan accordingly, especially around holidays and 3-day weekends. Are you working with live animals or people? You need to be prepared to have your project approved by an internal review board at your school site, a group of faculty who will look through your application to use living subjects (animals or humans) to ensure their safety and that they are being treated humanely. This review may take a few days or several weeks—be prepared to wait.

Allow time for final data collection and harvest, which can take large blocks of time. Recruit help for data collection but remember what goes on behind the scenes to ensure accuracy and precision. It is okay to have a helper record data for you as a secretary but you might want to consider taking all of the measurements yourself. Will you be able to use the facilities on the day you want them? Will you need permission to have access to scales, dryers, meters, or other pieces of equipment that may need to be checked out ahead of time? Now is the time to speak to lab technicians and other staff who can really help you get organized. Their experience with other researchers is vast and they will know of many shortcuts to help you along. Do not forget to say thank you when you are done with final harvest and data collection—you may need their services in the future!

Entering data into a computer spreadsheet is time consuming and somewhat tedious. As mentioned before, always back up your data (and lab report too!) in several locations (e.g., hard drive, flash drive, personal webpage, email accounts). There is nothing more devastating than realizing you have just lost the computer document you spent several hours composing. Backing up your work only takes a minute or two, and is the difference between a professional who takes her/his work seriously and an amateur, who tends to wing it at the last minute.

The next exercise is designed to help you learn how to read related journal articles to find information on a common topic. This is important because you will be assigned topics in your classes and will need to find out what is known about your topic.

I remember in grad school when I was conducting my literature review. It seemed that I was reading articles and books for months and months without an end in sight. One day my advisor stopped by my office to visit and check up on my progress. He asked how I was doing and what was I learning. I replied that it seemed that I was reading the same stuff over and over, and that made him smile. "Sounds like you're done reading for awhile," he said, knowing that when you see the same articles and authors repeatedly, you have probably exhausted your search.

I was able to move on to another study and was glad to have something new and refreshing to read! As for you, now is the time to take advantage of the new topics that you and your classmates will be sharing. Look through the next exercise and take time to read what each article is about. Use the shortcuts mentioned in this chapter to help you practice reading quickly but then go back and thoroughly read each article.

3 Planning the Experiment

Variables and Treatments

When you have decided on a topic to research (e.g., fertilizer brands for plants, milk replacer brands in livestock), it is critical to pinpoint what it is you want to test or observe. This is known in empirical science as the independent variable (and as the manipulated variable in the social sciences). The **independent variable** is the application or procedure that is being altered or changed in an experiment between groups of **experimental units** (the plants, animals, Petri dishes of bacteria, whatever is receiving the treatments). The independent variable's rate may be increased or decreased or its application may be compared to similar products to see if they are similar or different. The independent variable is described in the objectives in the introduction section of your lab report.

From here throughout this text, the term variety will be used to describe plant cultivars (cv) and varieties. Do you know the difference? Cultivars (cv for cultivated variety) are new plant types on the market that were discovered in a managed facility or landscape (e.g., on a farm, in a greenhouse, or out in nursery). A plant variety is technically a new plant type that was discovered in its natural habitat (e.g., the open range or forest). It is typical in the horticulture industry to call new types of plants "varieties" instead of cultivars, even though it may not be technically correct. Plant naming will be discussed further in class when we get to plant taxonomy in lecture.

Many of you are already familiar with plant varieties such as 'Jalapeno' pepper or 'Early Girl' tomato. The variety name is capitalized and surrounded by single quotation marks. In California, it is state law that all vegetable plants must be labeled according to variety. Would it not be disappointing to purchase one type of pepper thinking it is going to be hot and spicy but instead find out it is another type that is bland? It would even be worse if it were the other way around!

In livestock, species of animals are separated into breeds (what we typically call ethnicity or race in humans). A popular dairy cattle breed is the Holstein, known for their black and white markings and seen in clever milk commercials. Popular beef cattle breeds include Hereford and Angus, which has become popular in beef commercials. I tend to wonder though, are they sure they are selling steaks from Angus beef animals? How can they be sure? It does not matter to me though; I enjoy a good steak from any breed of beef cattle!

Regardless of whether you use animals or plants in your experiments, it is best to keep the independent variable manageable and somewhat small in scope. For our class, we will usually not go over five to eight rates of the independent variable to keep the experiment controllable.

The independent variable is broken down into rates or applications called **treatments**. Suppose we want to test some fertilizers used by homeowners to determine which product encourages more flowers in a landscape and is the most economical. We could use five different brands of fertilizer typically available from a home retail box store on a single plant species as the five treatments (e.g., Miracle-Gro®; Sta-Green®; Vigoro®; Bandini; and composted, bagged steer manure). The fertilizer type is the independent variable, the brands are the treatments applied to the flowers.

If we raised livestock, we could develop a small scale livestock experiment testing different brands of pig feed from different companies (Purina®, Nutrena®, Kent®, and Blue-Seal®). These feeds are treatments under the independent variable (feed type) and could be fed to feeder pigs to determine if the average daily gain of each pig varies due to feed type. In a garden experiment, we might conduct a test to study shortest harvest time in carrots. We could plant eight different varieties of carrot seed and measure the crop time from germination to harvest. Carrot variety would be the independent variable (the different types of carrots are what are being tested against each other); therefore each carrot variety would be a treatment.

An important point to make here is that there is a control group in most experiments. Experiments tend to have several groups of plants, animals, or other organisms. Each group would receive one of the treatments and all individuals in that group contribute data to their groups' mean. A **control group** is a group of plants (or livestock, fungi, bacteria, or whatever organism is being subjected to the treatment) that do not receive any of the treatments being tested. If the control group's individuals receive anything, they receive only plain water or the typical fertilizer (in the case of fertilizer experiment for plants) or the regular feed that has been typically provided (in the case of the feeder pigs feed trial). The use of a control group ensures there is a baseline value to compare the other treatments to. In the case of the fertilizer experiment mentioned previously, there would actually be six treatments, five fertilizers and a control treatment. For the feeder pig study, there would be five treatments, four feed brands and a control (typically the feed that has already been in use.) For the carrot experiment, the control treatment could be a standard, well-known carrot variety that is currently used in commercial production, and we could compare it to eight new carrot varieties that we want to put on the market.

Please note: in animal studies, care should always be taken to ensure all animals in experimental trials get the necessary feed and water they require at all times. To allow animals to go without water or feed is highly unethical and goes against all standards that scientists have against cruel and inhumane animal testing. No experiments of this type will ever be allowed in this class and all plants and animals will be cared for to the best of our ability. As mentioned previously, there are mechanisms in place at most universities to assist with animal or human subjects' testing and their approval must be given before any experimental treatments may be applied.

Another type of variable that we have alluded to but not directly mentioned is the dependent variable (also called the responding variable in the social sciences). The **dependent variable** is the outcome being measured. In plants these outcomes might include plant height, root length, dry shoot weight, flower number or fruit number and in livestock it might include average daily gain, pounds of milk, or percent backfat. Typically the dependent variable is quantitative—something numerical that can be measured objectively. In the social sciences involving humans for subjects, the responding variable can be subjective using survey data (which can change depending on a person's mood when they answer the questions) or can be used in ranked or categorical data. The research in this class will mainly focus on dependent variables that are quantitative but we will address qualitative data that might be important in agriculture and the food sciences (e.g., sensory testing, taste-testing, or visual appraisal of new food products or food packaging).

Remember that it takes hard work to narrow your topic down into objectives that are measurable. What are you trying to measure from your treatments? Is it shoot weight? Are you trying to increase the number of flowers or pounds of fruit yield? Are you hoping to increase crop yield or average daily gain in livestock? Are you keeping breeding ewes or does that have twins or triplets each time? Remember that the more focused you are in the beginning of the experiment (specifying outcomes you want to assess), the easier it will be to set up the details of your experiment later on.

In survey studies, response variable data are generally taken in the form of ranks. Many of us have seen Likert type scales during teacher evaluations. You score an item on an evaluation with numbers, for example, 1–5 with 1 being low and 5 rating high. Sometimes we see ranks of agree, neutral, or disagree or the answers put into categories (e.g., tomato fruit color: red, yellow, or orange) with the responses assigned a coding variable. **Coding variables** help with sorting categorical data by assigning a number to a category. From the previous example with tomato fruit colors, each color receives a number (e.g., red = 1, yellow = 2, and orange = 3). Having a number represent the color category instead of a descriptive word aids in counting responses and analyzing data. You just need to remember which number represented which color category!

We will discuss more of these types of analyses later on in the book. Remember, two important factors critical in setting up an experiment are deciding on the independent and dependent variables and these will be included in your experimental objectives.

Lastly, do not forget an often-overlooked factor: the controlled variables. The **controlled variables** are those environmental factors (e.g., plant type, greenhouse temperature, soil type, outside temperature, barn temperature, amount of water provided) or extraneous things that all of the treatments will be subjected to that have nothing to do with the experiment. Knowing what your controlled variables are improves consistency and stability in an experiment since you do not allow them to be changed or altered which can interfere with your experiment. Controlled variables are those things that you control as you set up the experiment (e.g., plant or animal selection, placement in a garden plot or feedlot, temperature) to ensure all treatments are being applied equally and without bias.

A word of caution here: do not get controlled variables confused with the control treatment. The control treatment is the baseline treatment found in almost all experiments to give the researcher something to compare the independent variables (other treatments) to. The point of experimentation is to determine if the experimental results come from treatment effects, not due to random chance (e.g., did the plants grow larger because they received more fertilizer or was it because the plants placed by the greenhouse wall received more sun, causing them to grow larger?). By using one species or variety of plant or a single breed of livestock in an experiment, you can reduce the variation between individual plants or animals. Will you eliminate the variation? No, but we are doing the best we can to reduce it and any interference (error or noise) it might cause.

For the feeder pig feed experiment, some controlled variables might 1) include having all of the pigs come from the same litter (hence, are the same age and have similar genetics), and 2) having the pigs fed the same amount of feed at the same time each day. For the plant fertilizer experiment, a beginning researcher should use only one variety of plant, the plants should be planted in the same brand of potting soil, in the same size of pot, grown in the same greenhouse, and given the same amounts of water each day. Consistency to all treatments (including the control group) helps to reduce variation (which causes experimental error). You want to do your best to make sure any differences in the dependent variable are caused by the treatments' effects and not by other factors! Otherwise, how would you be sure the outcomes were due to treatment effects? You would not—and that is what experimentation is all about, is there a difference due to treatment or due to random chance? We want to reduce all other factors and focus on the independent variable as the only source of variation. Then we should end up with a reliable data set (dependent variable) to analyze and interpret.

Try to recognize the independent, dependent, and controlled variables of a journal article by reading the materials and methods section. It is good practice to find each of these variables in this section and see how other researchers set up their experiments addressing each kind of variable. Reducing variation is crucial to ensuring a quality experiment and reliable results.

Reducing the Chance of Experimental Error

When we set up our experiments, we need to make sure we control all sources of variation to the best of our ability. This will allow the "true" results of the experiment to come through (not due to extraneous sources) and ensure that the results we found were due to treatment effects and not due to random chance or interference by environment. Variation can come from many different sources: performance from individual animals and plants, age, plant variety or animal breed, placement of **experimental units** (individual plants or animals) in groups, handlers' routines (e.g., did the plants or animals get watered the same way at the same time each day?) Reducing variation is key to setting up a good experiment.

Experimental error moves the true mean (numerical average) away from our data set and comes from many sources: one plant happens to be near the greenhouse wall therefore receiving more sunlight or one plant happens to be nearest to the overhead sprinkler, getting drowned at each irrigation, or ten different students weigh all of the feeder pigs on two different scales over a two-hour period. Will the first pig be weighed exactly in the same manner as the last pig in the group? Probably not, and anyone who knows anything about handling pigs knows they tend to lose weight when excited, moved around, and stressed out, especially over a long period of time.

Controlling Variability in Experimentation

There are several ways we will be able to control variability in our experiment and this section will show you how you can control accuracy, precision, replication, and randomization to improve your experiment's reliability. **Accuracy** does not need too much discussion; it is the condition of being true or exact. Accurate data are correct and best represent the actual or true answer. In our research, for example, we could use several fertilizer rates or brands to test which fertilizer treatment produces the best result (plant height, flower number, or other dependent variable that we agree upon ahead of time). Our data should represent each specific fertilizer brand's performance and should be a true representation of that product's performance. Will we ever be totally accurate? No, we are human, we will never have the ability to know exactly how much a plant weighs or how tall it is, but we can come very, very close. That is accuracy, getting as close to the true answer as possible. We do not have the means to be truly accurate but we will do the best we can.

We keep talking about using several plants or animals in our experiments, why is that? Why is it not a good idea to use only one plant per treatment? You might be wondering how many plants or animals or Petri dishes of fungi should be included in our experiment. Using one plant or one Petri dish of fungi would be inexpensive, with minimal inputs to manage, few experimental units to care for, and few to take data from. But that is the problem; there would be too few plants to take data from. Also, what would happen if our single test plant in a treatment died or if that single Petri dish became contaminated? The experiment would not necessarily be complete and have all of the treatments it should, therefore we probably would not learn as much since we could not speak about that particular treatment.

In reality, will the data from one plant tell us the true data that that rate or brand of fertilizer could provide? In other words, does one plant receiving that treatment represent an entire population receiving that treatment? Of course not, that is like saying a professional basketball player's height is a typical representation of the general population height of people in the United States. We all know that professional male basketball players' heights are not typical. They would not be a good single representation of the height of the typical American male, and therefore would probably be inaccurate in describing people's height in the United States. A better way would be to take the average (or mean) of a group of people throughout the nation to represent the mean national population height of Americans. That is a better estimate because you would find there are some shorter people, some taller people, and some in the middle included in the measurement. Taking the mean of a sample from the population is one of the best ways to gather data about a group of individuals; it includes all types of variation that might be included in that population.

Are you familiar with the concept of population and sample? A **population** is the number of individuals, plants, or animals in the entire living group. Sometimes you can take data from an entire population if it is small enough but usually you cannot (it may be too expensive to do so, the population may be too spread out over a large territory, or we do not know where all of the members of the population reside). When you cannot take measurements from each member of a population, we take some members from the population and that representative group is called a **sample**. What is important to note though, is that the sample size needs to be large enough to be representative of the population and there needs to be enough individuals in the experimental treatments to represent the variation that each individual from that sample has.

Like the male basketball player mentioned previously, his height is much greater than that of the general public but if you average his height in with those of 200 others selected at random, for example, from all over the United States, the extreme variation of his height would be reduced (diluted) since there would more than likely be shorter people in the sample as well. The sample size to take depends on the population number but it is best to have it large enough to provide several individuals per treatment (these are called replicates). Many horticulture researchers agree that a sample size of 30 individuals should be a minimum number to take from a population, with at least five replications/treatment. This would result in six treatments in this experiment. Are more replicates better? Certainly, but financial resources usually limit the number of replicates, particularly with livestock, which can be expensive to feed and maintain throughout the experiment.

Replications Within Treatments

Replicate number is related to the sample size required in your experiment. **Replicates** are the specific individuals assigned to each treatment and help reduce experimental error. In the calf milk replacer example, we might have five Jersey dairy calves each getting one type of milk replacer and testing for weight gain; therefore there are five replicates/treatment. If there are four types of milk replacer being tested in the study, we would need 20 Jersey dairy calves available for the experiment ($4 \times 5 = 20$). Do 20 Jersey calves in our study truly represent the entire population of all Jersey calves? We do not know for certain (and surely they do not), but we do know they are probably a better representation of the true weight gain that Jersey calves can exhibit than testing only one calf per treatment alone. Will all Jersey calves consume the same amounts of milk? No. Will all Jersey calves like the taste of the milk replacer? Probably not, so to account for this and other variations that individual calves have that you cannot foresee, several replicates need to be used in each treatment. This is why it is important to learn how numbers of replicates in an experimental design can affect our study's accuracy. It is better to start with more replicates than you think you will need (greater than five) in each treatment to account for any variation that may occur within the animals or plants assigned to that individual treatment. Five is a good number for beginning researchers in the animal sciences because it is obviously greater than one, is an odd number, will reduce expenses associated with caring for large numbers, and will statistically allow us to test for treatment differences.

For the Jersey dairy calf experiment, it would be recommended to have at least five calves all of the same age, same breed, and same sex equally distributed between all of the treatments. This would promote a valid comparison, and that is where the saying "comparing apples to apples, not apples to oranges" comes from. In the carrot variety experiment example where we were testing different varieties of carrots, seeds are easy to plant and generally inexpensive to handle and care for. Planting a dozen carrot seeds/treatment would be suggested, as long as each treatment was allotted the same number of seeds of each carrot variety.

To recap, to increase accuracy in an experiment, minimize variation, and detect treatment differences, more replicates/treatment need to be included. This will allow us to see if there are differences between each treatment due to the treatments' effects, and not due to random chance that they ended up that way.

Improving Experimental Accuracy and Precision

For this class, replicate measurements will be recorded in standard metric units (cm, kg, °C) and rounded to the tenths place (e.g., 10.9 cm, 27.9 °C, and 0.6 kg). Means within each treatment will be calculated from all individuals in the treatment and rounded to the tenths place as well. We are going to spend a lot of time taking initial and final data in lab groups made from your classmates so it is very important to discuss measurements and how precision can increase accuracy.

Measuring some object is a subjective activity that no one does exactly the same way every time. A way of introducing error to an experiment is through taking careless measurements, which is a lack of precision. **Precision** is the ability of a measurement to be consistently reproduced. It is like doing something the same way every time. If every member of the class were given a paper cup and told to measure it with their own ruler, there might be almost 30 different measurements! Without specific instructions, some students in the class would measure the diameter of the opening at the rim, some would measure the diameter of the bottom, some students would measure the length of the cup along the side, while others might measure the cup standing on end from table to cup rim.

I hope a few things are obvious here: first, the directions of how to take the measurement were not given so everyone thought what they were doing was best. Second, did everyone measure only once or did they take another measurement to verify they were doing it the same each time (precision)? Third, were the students taking and recording their data in metric units (cm)?

Let us measure the cups again, this time along the side of the cup from the bottom edge to the rim. Everyone would know to place the ruler against the cup and take the measurement in this manner.

Everyone's measurement would be more precise which in turn would probably increase accuracy. If the cups (and rulers) are all the same brand and from the same package, they should almost all have the same measurement.

Let us look at how precision affects your life right now. You know how to be a good student, right? You go home or to the library after class, re-reading your notes and textbook. You confirm that you understand the concepts and take the reading quizzes. You study the homework and the correct answers and check with your instructor during office hours if there is anything you are uncertain of, right? You look at the suggested websites to reaffirm concepts given in class. If you do all of these things, you are probably performing well and earning good grades on the midterm exams, right? If you are accurate, you got the information right on the exam and earned a perfect score (100% accurate I hope!)

What contributes to your accuracy (and therefore, good grades)? Your precise study habits, that is what! You are consistently reproducing the same tasks every week. Every day without fail you schedule time in between your job and other activities to review the daily lecture notes and do the assigned reading; you arrange to take the weekly reading quizzes before the weekend and read the lab procedures in advance of lab; you ask questions in class and review the notes when you get home. You study the same way, consistently, each and every week. This consistency in study habits, doing the same things the same way is precision. You pay attention to the course details by studying what is coming up in the syllabus and it pays off when you earn good grades. You have prepared consistently since you see what is working to help you learn.

Does precision always lead to accuracy? Of course not, maybe you have a friend, Earl, who does not take his class work all too seriously. Earl rarely attends his classes and when he does he sleeps in the back row since he stayed up late the night before. On weekends Earl stays out late partying, hangs out with his friends playing videogames, sometimes he plays sports in the afternoons and then goes to his part-time job, only to find he has to squeeze in time to take the reading quizzes late Sunday night because he put them off during the week. Earl is behind on the reading assignments and ends up cramming all night before the midterm, which is something you do not do because you have been studying the class material all along.

Earl is not dumb but tends to earn bad grades in most of his classes and cannot understand why he is doing so poorly in school. But you know why; he is very precise in his poor study habits. Earl does the same things each week that take him away from studying (partying, goofing off, etc.) and ends up with poor grades. Earl fails the midterm due to his lack of studying (his answers were <60% accurate) but wants to improve, his college degree depends on it! How would you suggest Earl change his study habits to improve his class performance before it is too late? Do you think that changing just a couple of things would make a difference? I bet it would too.

Can you now distinguish accuracy from precision? Can you see that it is crucial to the success of the experiment to make sure you use the same methods, use the same procedure, the same equipment, the same cuttings or seeds, and the same measurement techniques each time? Getting everyone in class (or in your experimental groups) to make taking their measurements and planting techniques seriously is key to ending up with data that are reliable and "true." And that is all we're asking you to do, do the best you can to ensure that our final data are as true of a representation that they can be.

Randomizing Replicates Into Treatments

Some variables can be controlled in an experiment, some cannot. A clever saying about variables found in experiments is "block what you can, randomize what you cannot." Remember when we were discussing controlled variables? We kept the environment, water, temperature, and other factors that were not being measured the same with all treatments. That was like blocking. We will discuss blocking experimental units more in detail later on.

One thing we can control is placement of experimental units to the treatments. In other words, we need to make sure that every plant or animal (whatever our experimental units are) has the same chance of being put into each treatment. Why is this important? Why should we care which plants go into each treatment in the experiment? Randomizing experimental units gives each one the opportunity to be placed in all treatments, which reduces the chance of introducing bias. Here is an example: you are testing different levels of fertilizers on 'Roma' tomatoes to determine if there are any differences between their fruit production. When you transplant the **plugs** (single seedlings grown in trays) into pots, you may notice some of the plugs are smaller than the others and think to yourself, "Well, these plugs are smaller so I'll just put them in the treatment groups that will be receiving lower rates or no fertilizer at all since they'll probably perform worse anyway." What's wrong by making this assumption and putting weaker plants in lower-quantity treatments? You are introducing bias into the experiment. Should not all treatments have the opportunity to perform at their best? They should, and you need to do everything that you can to reduce or eliminate bias—intentional or not. Set the small plugs aside and select from the most typical, lush plugs in the tray. Extra plugs can be potted up and either given to staff members or grown out and planted around the buildings or out in the landscape.

It is easy to find seedlings or rooted cuttings in horticulture experiments that are consistent in size and weight but it is harder to do that with livestock. Animals need to be weighed and could be separated by gender. Different breeds of livestock may need to be taken into account: do some breeds grow faster or differently than others? If you have a mix of gender and breeds in livestock experiments, it is a good idea to assign each of the sexes to each of the treatments, that way each treatment is represented by both gender and breeds. This method of separation is called blocking. **Blocking** helps remove extraneous variation that can cause noise (error) in an experiment. Remember, we want to distinguish treatment effects and only those effects, nothing else. We need to minimize variation as much as possible so blocking into similar groups that have each of the breeds or genders reduces variation. With blocking, we would be able to make treatment comparisons between each group (or block) that had the different breeds (or gender) within each block. Putting all one gender into a single treatment confounds the issue, did the animals gain what they did due to treatment effects or was it because their gender gave them an advantage?

In plants, we could test products on different plant types (poinsettias, lilies, houseplants, and peppers for example) and could place several specimens of each plant type into each treatment. Again, we would block to remove variation by plant type. Blocking would allow us to see if there were treatment differences across all plant types since all of the plant types are receiving each type of treatment. Can you see why it would be incorrect to give only one treatment to a group of poinsettias, another treatment to a group of lilies, a third treatment to some foliage houseplants and a fourth treatment to some vegetables? We would be comparing apples to oranges. Lilies grow much differently than peppers so we could not tell if one performed better due to treatment effects or because the plant type was a faster, more robust grower.

By now you have probably realized there is a lot to consider when randomizing plants or animals into treatments. Do you know how you are going to label the plants or animals once you select which ones are going into the specific treatments? Working with bacteria or fungi is easy; you simply label the Petri dish. But what do you do about livestock and rows of corn in a field? Do you have plant labels or ear tags or know the animals' specific markings (tags, brands, or tattoos)? Are you randomizing the experimental units (individual plants or animals or rows of corn) and placing them in their final location (e.g., steers taken to a specific feedlot) immediately or do they need to be moved to another location later on (e.g., plants to a greenhouse bench to a nursery)?

In livestock, we might have to take weight and gender into account and block to reduce bias. In plants though, we have the advantage of growing them from seed as **F1 hybrids** or **vegetative cuttings** (clones). Test plants could all be **clones** (genetically identical) so it would not matter which plant went into which treatment. That is a definite advantage of using plants over animals in beginning experimental research and why the author is a bit biased toward them (besides, some vegetable varieties tend to grow really fast and you get to eat the mistakes!).

Once you have the experimental units that you are going to use in your study selected and separated away from the rest, take them and randomly assign them to the individual treatments.

This can be done by a lottery system with a random number generator on a spreadsheet program (such as Excel) or accomplished by hand through a bingo-style lottery selection system. To put plants into a fertilizer trial, pot up the number of plants you will need to provide the appropriate number of replicates/treatment. Using plastic pot labels and a permanent marker or a pencil (so the information cannot be washed off), record the treatment information and plant number on each label. Plant label manufacturers make 5" or 6" labels in several colors (pink, green, blue, yellow, red, orange, and lavender) in addition to white. You could use different colors to represent your treatments so you could quickly find the plants that belong to a specific treatment on the greenhouse bench or in a nursery. This also facilitates taking measurements from the treatments and aids in plant identification at a later date.

With livestock, use of ear tags serves the same purpose. Many animals already have ear tags or notches for identification so those already in place on an animal can be used; just remember to keep accurate records of what each identifying marker means and what treatment or animal number it represents.

To randomly assign the plants that you want to include in the experiment into treatments, place the plant labels (already filled out ahead of time) in a bag and randomly draw each one, placing it in a potted plant (randomly taken from the bench). Repeat this step until you have drawn and placed all of the plant labels, checking to make sure you have the correct number of replicates/treatment. With animals, you can have a list of the animals' ear tag numbers and their corresponding treatment codes listed on slips of paper and mixed up in a bag. The random draw from the bag allows each experimental unit to be randomized into treatments; every plant or animal had an equal opportunity to be in each treatment. Use of a random number generator on Excel assures that each experimental unit has the opportunity to be assigned truly at random. Do you know how to use the random number generator? On Excel 2003, go to the menu Insert, Function, RAND. Highlight the column where you want the random numbers to appear. Turn these random numbers into whole numbers by changing the function to =RAND()*100 by adding the *100 to the end of the function equation. You can have your list of replicates in one column and the random numbers in another column and can sort by the random number (Figure 3-1), putting them into the order of placement on the bench, in the nursery, or wherever the final destination is.

Ask yourself this, when assigning experimental units to treatments, why would you not simply take each plant or animal that was nearest you (on the bench or in the corral) and place it into a treatment? Is it possible that the animals closest to you could be the friendliest (or oldest)? For plants, could the plants taken from one end of the bench be the smallest due to where they were located in the plug tray? That is, could the smaller plants be transplanted together and end up being grouped next to each other? Could all of the slower cattle be selected first? Could this be a problem later on? We cannot answer that and do not want to have to think about that, so it is best to randomize all units into our treatments and forget about it. We did the best we could to minimize experimental error by randomizing, and that is all we can do.

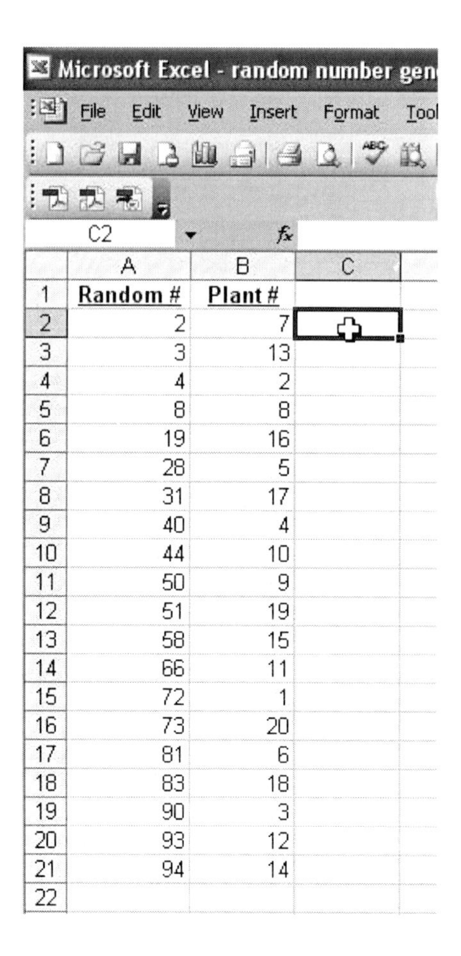

	A	B	C
1	Random #	Plant #	
2	2	7	
3	3	13	
4	4	2	
5	8	8	
6	19	16	
7	28	5	
8	31	17	
9	40	4	
10	44	10	
11	50	9	
12	51	19	
13	58	15	
14	66	11	
15	72	1	
16	73	20	
17	81	6	
18	83	18	
19	90	3	
20	93	12	
21	94	14	
22			

Figure 3-1. Assigning a random number to each replicate in each treatment on Excel aids in randomizing plants in an experiment and eliminating bias.

What is the best method to randomize units into treatments for small scale experiments? First of all, you need to know what your treatments are, so list them in your notebook. You could always use the random number generator but while it is easy, it is not very fun. To practice randomizing a small number of plants (or animals in a livestock experiment) into treatments using a bingo or lottery method, your instructor will provide your group with any of the following numbered items: deck of cards, ping-pong balls, poker chips, checkers, dominoes, or any other small objects that can be numbered individually. These numbered objects will be assigned to the treatment replicates and the random draw of each object will determine placement of the replicate into the experiment. Make sure the items you use are all of the same size and weight so they have an equal chance of being selected from a hat, box, or bag.

In your notebook or on Excel, give each treatment a coding variable (Table 3-1). This coding variable will help us keep track of each replicate that belongs to each treatment, which in this example is testing several rates of nitrogen fertilizer with units of parts per million (ppm N). Table 3-1 gives the coding variables and individual replicate numbers for the plants used in our example fertilizer study which would be found on each plant label in each pot. Each replicate is numbered to distinguish one replicate within a treatment from another when we collect individual plant data throughout the experiment.

In Table 3-1, there are five replicates/treatment and each replicate is assigned a number between one and five. There are 7 treatments, including the control (no fertilizer, 0 ppm N), totaling 35 plants needed for this experiment. Should you only plant 35 plants from seeds or cuttings? It is a good rule of thumb to

Table 3-1. Assigning a numbered playing card to each replicate in each treatment aids in randomizing plants in a fertilizer experiment (rates of N) on a greenhouse bench.

Treatment Rate/Coding Variable	Treatment/Replicate Number	Card Number
0 ppm N / 1	1-1	1
0 ppm N / 1	1-2	2
0 ppm N / 1	1-3	3
0 ppm N / 1	1-4	4
0 ppm N / 1	1-5	5
50 ppm N / 2	2-1	6
50 ppm N / 2	2-2	7
50 ppm N / 2	2-3	8
50 ppm N / 2	2-4	9
50 ppm N / 2	2-5	10
100 ppm N / 3	3-1	11
100 ppm N / 3	3-2	12
100 ppm N / 3	3-3	13
100 ppm N / 3	3-4	14
100 ppm N / 3	3-5	15
150 ppm N / 4	4-1	16
150 ppm N / 4	4-2	17
150 ppm N / 4	4-3	18
150 ppm N / 4	4-4	19
150 ppm N / 4	4-5	20
200 ppm N / 5	5-1	21
200 ppm N / 5	5-2	22
200 ppm N / 5	5-3	23
200 ppm N / 5	5-4	24
200 ppm N / 5	5-5	25
400 ppm N / 6	6-1	26
400 ppm N / 6	6-2	27
400 ppm N / 6	6-3	28
400 ppm N / 6	6-4	29
400 ppm N / 6	6-5	30
800 ppm N / 7	7-1	31
800 ppm N / 7	7-2	32
800 ppm N / 7	7-3	33
800 ppm N / 7	7-4	34
800 ppm N / 7	7-5	35

over plant by 10–20% to account for slow growers, sick plants, or poor performance. After a few days from transplant, select your experimental plants from that group that you will use. Extras can be given away to colleagues, used for other experiments, or used to decorate your office or lab. Do not underestimate the power of giving away houseplants to staff members and planting out in the landscape!

To randomize the replicates that you grew for this fertilizer experiment on a greenhouse bench, for example, we need a method of selecting each replicate at random to place into each treatment and on the bench. This is easy to do using a deck of playing cards for our example: take a large permanent marker and number each card 1–52 (if it is a regular deck of playing cards) and assign that number to each of the replicates (Table 3-1). When that card is drawn, that replicate is selected and placed on the bench in the order drawn. For example, if card #10 from Table 3-1 is drawn first, replicate 2-5 is placed first on the bench. Determine the direction (top to bottom, left to right) ahead of time that the replicates are going to be placed on the bench (Figure 3-2). Shuffle the numbered deck of cards and draw each card in the order to place the plant, selecting the plant number of the card (with corresponding replicate from your list, Table 3-1) that goes in that location (Figure 3-2). From the random number generator from Excel (Figure 3-1), we sorted the random numbers and their plant numbers to get the order of placement on the greenhouse bench. Same idea, only it uses a different method.

Since we need to randomize 35 plants, set cards numbered 36–52 aside, we will not be using them right now. Thoroughly shuffle the remaining 35 cards to randomize them. As each card is drawn, the plant that corresponds with this card number is found from the table and placed on the bench where it will remain for the duration of the experiment. Even if the plant dies and is removed from the bench, its space is left vacant. Again, this reduces experimental error by reducing the chance of having all plants in one treatment together near a drafty greenhouse door, under a leaky glass roof panel, near a sprinkler, etc.

Figure 3-3 and Figure 3-4 show how the numbered playing cards were drawn and placed in the predetermined order on the greenhouse bench. Students can work in teams to lay out the cards and place the corresponding plant that goes at each location in a "scavenger hunt" manner.

It is a good idea to look over the final placement of the plants and see the "mix" of replicates. Using colored plant labels makes it easy to find replicates when data needs to be collected or more treatment needs to be applied. Please realize that although we did our best to place plants randomly, sometimes several plants in one treatment end up together in the same area regardless of our attempts to eliminate bias. In Figure 3-4, notice treatment 3 has several of its replicates in the lower right-hand corner of the bench, and that is okay, we did the best we could to randomize them and

(Top, north edge of bench)

START	x →	x	x	x	x	x
Place plants from left to right as their number is drawn from the deck of cards →						
	x →	x	x	x	x	x
Continue placing plants left to right, then continue with the next row down →						
	x	x	x	x	x	x
Continue placing plants left to right, continue with the next row down →						
	x	x	x	x	x	x
Continue placing plants left to right, continue with the next row down →						
	x	x	x	x	x	x END
Continue placing plants left to right; finish with last card and last plant →						

(Bottom, south edge of bench)

Figure 3-2. Experimental plant placement on a greenhouse bench. Orientation of plants was predetermined to be placed on the bench from left to right, top to bottom. In this example the top row is on the north wall.

(Top, north edge of bench)

29 4 16 7 8 23

Playing card #29 was drawn from the deck first, followed by cards #4, 16, 7, 8, and 23 →

2 14 28 5 30 9

The card for plant #2 was drawn next, followed by cards #14, 28, 5, 30 and 9 →

10 1 17 26 24 15

Playing card #10 was drawn next, followed by #1, 17, 26, 24, and 15 →

21 6 3 19 12 18

Continue placing cards left to right on the bench as their number is drawn from the deck →

27 20 25 22 11 13

Finish the last row with the last 6 cards, ending with card #13 →

(Bottom, south edge of bench)

Figure 3-3. Numbered playing cards were randomly placed on a greenhouse bench after being drawn from a shuffled deck of cards marked 1–30. Orientation of placement was predetermined as left to right, top to bottom.

(Top, north edge of bench)

6–4 1–4 4–1 2–2 2–3 5–3

The 1st plant drawn from the deck was in treatment 6, replicate 4 →

1–2 3–4 6–3 1–5 6–5 2–4

The 7th plant drawn from the deck was in treatment 1, replicate 2 →

2–5 1–1 4–2 6–1 5–4 3–5

The 13th plant drawn from the deck was in treatment 2, replicate 5 →

5–1 2–1 1–3 4–4 3–2 4–3

The 19th plant drawn from the deck was in treatment 5, replicate 1 →

6–2 4–5 5–5 5–2 3–1 3–3

The 25th plant drawn from the deck was treatment 6, replicate 2; the 30th was treatment 3, plant 3

(Bottom, south edge of bench)

Figure 3-4. Placement of each replicate on a bench that corresponds to the card drawn from the deck. Plants will remain in this location for the duration of the experiment.

they were placed randomly. Sometimes experimental units from the same treatment end up being placed adjacent to each other. That is the way it goes.

By mixing up all of the treatments together in one big area, this is an example of a **completely randomized design (CRD)**. Each treatment is applied to the same type of experimental unit (same variety of plant or animal breed). We will discuss other types of experimental designs found in agriculture and biology in the next chapter.

Do you have to use numbered playing cards to randomize plants in an experiment? Certainly not, you can number and use anything that could be selected in a lottery-type style that can be numbered and selected randomly from any container, or you can use Excel to generate random numbers quickly. What matters most is that if you use cards or chips the pieces all have to be the same size and have the same chance of being selected compared to the other pieces. This is an unbiased method that lessens the chance for experimental error. If we have small experimental error, we have a greater chance of determining that any treatment differences came from the treatments themselves and not something else.

What About Larger Experiments Using Two or More Plant Varieties?

For some experiments you might want to compare two varieties of one plant species within each treatment (maybe you want to see if the varieties respond similarly to your treatments). You can differentiate the plants easily in the greenhouse (since they might look similar when they are young) by using two different colored labels (e.g., pink pot labels for geranium 'Showcase Rose Flair' and blue pot labels for geranium 'Fantasia Neon Rose'). Label each variety with a treatment code and replicate number across the top of the label. For example, the first replicate of 'Showcase Rose Flair' treatment 1 could be made on a pink label with a 1–1 (written in permanent ink) across the top. The first number 1 on the label is the treatment group (control group for example) and the second number 1 is the replicate number for this variety. This plant would be known as 'Showcase Rose Flair' control group, plant number 1. This is similar to ear tagging in swine production (litter number, pig number) and uses the fertilizer coding variables given earlier in Table 3-1. Figure 3-5 shows how we will fill out our labels here in this class for lab activities. The label on the left includes the date across the top, the experiment number (this label belongs to the soilless media experiment, 7-4), and gives the treatment (this plant would be placed in the control group treatment). On the back of this label (not shown) are the plant variety and the initials of all students in that group. Students should also specifically keep track of their treatment groups in their lab notebooks.

The label on the right in Figure 3-5 is typical for experimental research in hortscience. The date across the top is written differently, it does not have any slashes or dashes between the numbers. Writing the date in this manner also tells the viewer what experiment this belongs to.

Rarely will two or more students in a lab start their experiments on the same day so it is handy for the supervisor or faculty advisor to have a list of experiments by their number and who they belong to. Data sheets, figures, photos, tables, and other information can have this specific number in their title and it helps keep the experiment running smoothly and organized. The 1–4 written below the date on the label on the right of Figure 3-5 is that plant's coding variable. This individual replicate is the fourth plant of the first treatment. If this were the control group, this plant would be the fourth control plant found in the study. Again, using colored labels would help simplify this even further, having the control treatment marked with white labels, another treatment marked with orange labels, and yet another treatment having yellow labels would help speed up the search and measure process. Stay away from using dark-colored labels in your plants, they may be hard to find (silver or tan can blend in with the plant). Bright colors include light blue, orange, red, white, yellow, pink, and lavender. Sometimes green is hard to see too.

Another quick tip, while you are making up your labels during lab set-up or during your experiment when you have some "down time" and are just collecting data, now is a good time to label your paper bags that you may be using to take dried weights. Dry shoot weight is a common dependent variable measurement that is taken in horticulture, crop science, and biological studies so having your bags labeled for each individual replicate well ahead of time is a good idea.

When you look at a paper bag, you will see there is the side that has the bottom folded up facing you. Write on this side of the bag with large numbers using a large permanent black marker. Do not write

Figure 3-5. Use of plant labels helps with research plant identification. The label on the left is an example of labeling done on in-class experiments. The label on the right is typical of research experiments.

at the top; come down a few inches, so when you fold the top over and staple it your coding variable/plant number will not be hidden from view. The reason you should write on the folded side of the bag facing up is so that you can grab the bag quickly by the fold and open it and it also makes the bags easier to separate when you go to grab one during harvest.

In summary, experiment planning involves many things related to your objectives, from determining independent and dependent variables, reducing experimental error by controlling variation, determining appropriate sample size and replicate numbers, and randomizing experimental units into treatments. You have learned there are many things to consider prior to beginning the actual experiment, and I have tried to make it as simple as possible! Take the time to plan each stage of your experiment (including harvest) and it will save you a lot of time (or heartbreak) at the end. You will end up with more accurate results which will lead to a more thorough lab report.

Complete the next exercise using the journal articles selected by your instructor. Take some time to identify the variables in each study and know what each one contributes to the experiment. Categorizing the variables will help you get organized and meet the objectives of your research.

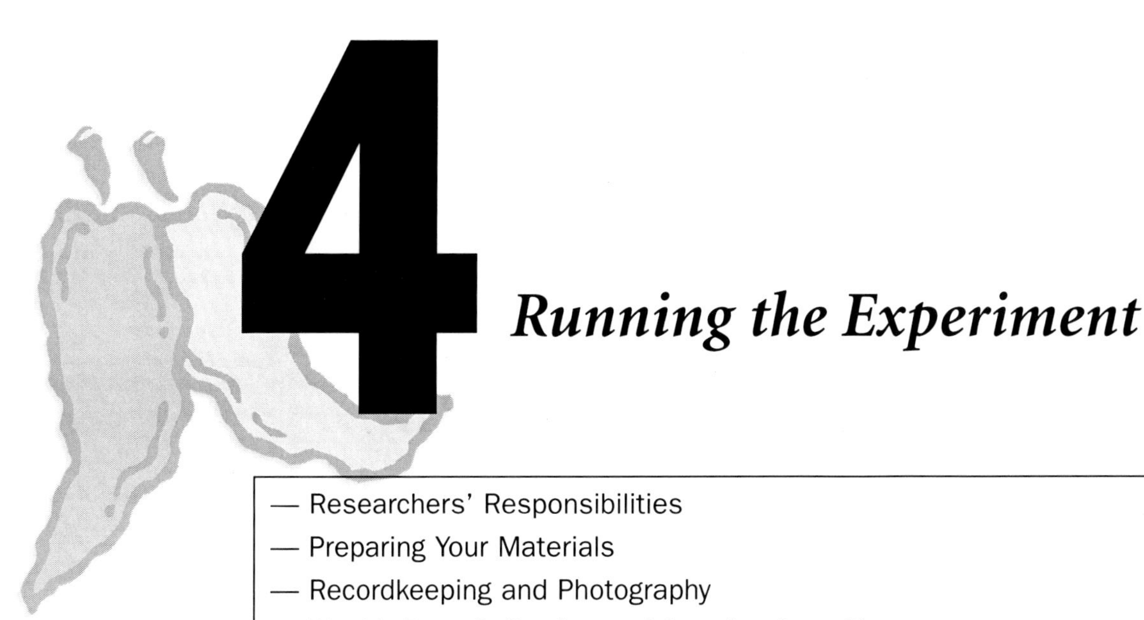

4 Running the Experiment

Researchers' Responsibilities

Now that you have determined your objectives, you need to start thinking about the daily or weekly maintenance of your experiment. How are you going to care for your experimental plants or animals? Who is going to provide for them, water them, feed them, and measure their growth or output? How will this be done accurately and with precision? Who will be called in case of an emergency and what is the back-up plan if the experiment fails in the beginning?

These and so many other questions need to be considered before any plants are grown or animals are purchased. As a caregiver of living organisms (and yes, bacteria and yeasts used in research count in this regard), you have a duty to those individuals to give them the utmost of care as you would a pet or family member. Under no circumstances will any animals be denied feed or water and they should always be provided warm, clean bedding and shelter from the elements. Plants will always be given ample water and will not be left untended or be allowed to become infested with pests or weeds. Responsibility is a trademark of a serious professional researcher and when research involves living experimental units, it takes on a greater sense of commitment.

Sometimes there is a lab technician or staff member who takes charge of caring for experimental units for classes on weekends or holidays but for advanced students (senior research project or graduate students working towards a Master of Science or PhD) the responsibility falls on that individual's shoulders. These are advanced research degrees and so part of the degree involves caring for one or more experiments. Usually the faculty mentor does not take an active role in the research but stays "on the sidelines" as an advisor or coach. The undergrad or grad student is ultimately responsible for the project and its success. In this class, there are several key staff members in addition to the instructor who will care for your experiments after hours and on weekends but you need to do your part by ensuring that the plants or other experimental units are set up for easy care: the irrigation is left on at each garden site, weeds are pulled and removed from the garden area or greenhouse, plants are checked for pest infestation, the work area is maintained and kept clean, and tools, equipment, and materials are stored and put away in the correct location at the end of each lab session. Informing the instructor of broken or missing lab equipment is your duty and you owe it to the next group of students following your lab to let someone know there is a problem. What may not seem like a problem to you may indeed become a safety hazard to someone else. In the gardens, the irrigation is a solid set system so make sure to tell your instructor immediately if you see any leaks or broken valves. These must get repaired before the next irrigation.

Sorting out the experiment by running through it in your mind and describing it on paper will help you think through potential problems or risks. Work with your lab partners and review the materials and methods sections of relevant literature to see if there is anything you might have missed. Communication with your instructor and lab partners is vital to running a successful experiment and you should always make sure you have your instructor's approval before beginning any experimental procedures. If you are working with human subjects, you must obtain approval from the institutional review board or human subjects review committee at your particular school site. This review ensures the proper care of living animals and individuals involved with your research study.

Preparing Your Materials

Take some time to go over your list of materials and supplies needed to run your experiment. Do you have enough pots, media, labels, fertilizer, plants, and bench space? Where are they coming from, who ordered them, and when will they be arriving? For livestock experiments, who is providing the animals, when will you pick them up or are they being delivered (and to what location), what will they be fed and how will they be watered, how will they be sorted into treatments, and who will be collecting the data? What are you going to do if your plants become infested with pests or weeds, will you spray them with insecticide, incorporate systemic pesticides, or release biological controls? It is okay to apply pesticides as long as all treatments get sprayed; it has to become a controlled variable. Besides, aphids or other small insects will quickly move to another host plant when one has been sprayed or drenched with insecticides so it is better to treat them all at once.

Take time to see for yourself that your research materials are located where they are supposed to be: that the pots are in fact in the storage shed, that there is enough straw or bales of shavings in the barn for bedding, and that the irrigation and water and electrical outlets work. Do you have to use any data loggers or other technology to monitor temperature, relative humidity (RH), light, or other data? Test these in advance to ensure they are working properly.

Data loggers and recorders come in many shapes and sizes and can record environmental conditions to be downloaded at a later date. Did you check to make sure these and other measurement devices (pH and EC meters, scales, etc.) have fresh batteries or are they able to run on AC power? Take some time to look into purchasing good equipment and have it on hand if you are going to keep working in the agriscience research arena. The high-end, sophisticated data logging equipment will pay for itself in the long run. Many data loggers simply run on small calculator-style batteries and are placed in the greenhouse or barn off to the side, out of the weather. They can be programmed to collect data as frequently as every 5 seconds or delayed to every hour or more. Be aware that more frequent collections by the unit will consume its memory or use more battery power so the logger will need to be checked often to ensure it is still working and collecting data. The data can be downloaded onto a laptop computer on the spot and reset to gather data once again. There are many brands and companies on the market so look for those that have products that are easy to launch and download (Onset Computer Corporation and Spectrum Technologies Inc. have good products that are solid for academic use and commercial use, check out their websites). There are many good products and companies with these types of technology, conduct a search on the WWW to find out more information.

Remember the old saying: an ounce of prevention is worth a pound of cure. Being prepared well ahead of time will save you hours of headache and heartache later on. Make sure you keep your research advisor updated on your progress and ask for help when you feel lost or overwhelmed. Do not let small problems or concerns fester until they become big ones; nip them in the bud, and your advisor or instructor can help ward off potential problems.

In case there is a crisis or emergency at your research site, do you have the critical contact information for staff members or emergency personnel to help out? Are there fire extinguishers and hoses available in the lab, greenhouse, or in the barn? Who do you call in case you cannot make it to your shift to care for your experiment, particularly if it involves animals? Having a contact list through a class website allows for easy email requests or discussion chats to help arrange carpools, study groups, and data collection teams. The technology is there to assist with teamwork, use it!

Recordkeeping and Photography

Speaking of teamwork, sometimes it is a good idea to make sure everyone in your lab group has copies of all of the data at all times. You can post current data for your lab partners on class websites or email copies of the weekly data to each other. There is nothing more frustrating to a researcher than losing data or information and with today's availability of email and text messages through the Internet and cell phones, it should not happen, though unfortunately it does. Hard drives crash, laptops get stolen, and notebooks get lost. So do not risk it, back up your computer's hard drive often. Do not take chances because it will happen to you, maybe not today, maybe not tomorrow, but you will have a computer crash someday. It might be as simple as losing your flash drive or having your laptop get wet from a spilled drink, but somewhere along the line you will lose a file, and it has happened to everyone at one time or another.

Backing up your work can be as simple as having an external hard drive, flash drive, burning a CD or DVD disk, or emailing a copy of the document to yourself to save on another computer. Whatever your method is, use it and use it constantly. Sometimes your instructors will have space available on their class websites for data storage for the entire class to access. Ask them for details. Regardless of the method, back up your files consistently and often, take no chances in losing your work, pictures, or research papers. If you do lose your data, sometimes it can be recovered but sometimes it cannot. It can be costly to have a professional computer engineer work to recover your data and it will take a lot of time to reenter data from hard (paper) copies so take an extra minute to back up your work.

Another way of keeping track of your experiments' progress is through digital photography. Taking digital photos for projects and presentations is not as expensive or difficult as it used to be when all we had available was print film. Color images really liven up publications and presentations and emphasize the message you are trying to project to your audience and with the availability of digital cameras, taking photos should be easy.

It may sound like an obvious suggestion but have a classmate or friend take your picture doing different procedures throughout the duration of your project. If you are designing a horticulture experiment, have a friend take pictures of you sowing seed, transplanting your plants, watering, mixing fertilizers (wear safety glasses or other protection!), measuring, recording data, and finally, harvesting. If you are designing a livestock study, have your friend take your picture with your animals, giving injections, weighing, feeding, applying treatments, ear-tagging, attending to parturition, bottle-feeding, terminal harvest (if applicable), or whatever it is that is relevant to your study. Do not look right at the camera with a big grin but act engaged in the activity you are doing. You can be serious in your pictures but look as though you are enjoying the activity you are doing.

A final thought about action photos, make sure you are not wearing the same shirt, hat, or clothes in all of your photos. Even if you stage your photos, you do not want them to look too phony or not genuine.

When you are getting ready for terminal harvest, make sure you are prepared by having an extra camera and extra batteries available (those of you using rechargeable batteries should have the charger handy) and an extra memory card. A lot of researchers have a laptop computer available so they can download their images and see immediately if they turned out how they wanted. It is a sick feeling to terminate your plants at final harvest and then find out your final harvest pictures are blurry or fuzzy when it is too late to retake them.

How do you take good photos for experimental presentations and publications? That is easy, remember to think about what you want your audience to see/know from the image and go from there. When you look at images in relevant publications (called figures in most journals) you will see many overhead shots and side-by-side views of example plants or animals from each treatment. Most are labeled by treatment in the photo and again described in an accompanying caption (either placed above or below the figure, see your particular journal for specifics on its requirements). Either way, you have to tell the reader which treatment is which.

For successful experimental photography, select the most typical of your experimental units. Do not choose the tallest, largest, or smallest plant from each treatment group; select the most average of the replicates, the one that best represents that treatment. The same thing goes for livestock or biological experiments. Those replicates at the extreme measurements might reflect a typo or simply may be just a fluke. In science, these extreme measurements are called **outliers**, and they are data that are not typical when compared to the rest in the treatment.

Outliers are easy to ignore when selecting animals or plants or Petri dishes for photography but not in data analysis; we will discuss that later. Consider how long it will take to get your experimental units to the photography site, and build that into your timeline.

For horticulture studies, set up the photography area in a cool or shady location. This will help prevent wilted plants. A classroom with good lighting is best but you might be restricted to late afternoon or evening use since it takes a couple of hours to take pictures and you cannot really be disturbed once you begin. Setting up the shot is critical and most of the set-up time is dedicated to making sure the experimental units are in the right place, look good against the backdrop, etc. Most photo shoots occur the day before terminal harvest since harvest is so busy and hectic. It is best to have the pictures, look at them, and then file them and back them up. Terminal harvest will take several hours depending on how many dependent variables are being measured.

Figure 4-1 shows how the author set up an overhead photo of treatment representative Vinca 'Pacifica Red' plants (*Catharanthus roseus*) by treatment (mM P) in this fertilizer trial. The plants were placed on white butcher paper on the floor (as the backdrop) with the experimental plants and identification cards placed on the paper. The tripod and camera are stationed in a sturdy location (although the tripod in Figure 4-1 does not look too sturdy) and was accessed by the author standing on the step ladder, looking down over the top of the plants.

Make sure you can move desks, chairs, and tables out of the way to make room to set up your camera and backdrop but remember to return them to their original locations when you are finished. No one likes to teach in a disorganized classroom so do your part and clean up after yourself. Be sure all of your plants are well-watered ahead of time so you do not have wilting specimens and have to delay

Figure 4-1. Taking an overhead shot of Vinca 'Pacifica Red' research plants involves a tripod for steady, clear photos.

shooting or worse yet, have your plants dripping water on your backdrop panel. Figure 4-2 shows what the author was seeing through the camera viewfinder during the overhead shot. Cropping and moving the picture with photo editing software will result in a straight photo with all treatments represented. Make sure you have your plants and labels straight (which they are in Figure 4-2); the rest can be edited. If you do not want to use printed labels, insert the treatment names with photo editing software or presentation software. The latter gives you access to adding text to a figure in a slide.

When taking photos of research animals, make sure they are not disturbed prior to the shoot, let them rest and have access to feed and water so they do not become restless during the photo shoot. Animals will be nervous as it is; do not add to their discomfort of being in a new environment by keeping them

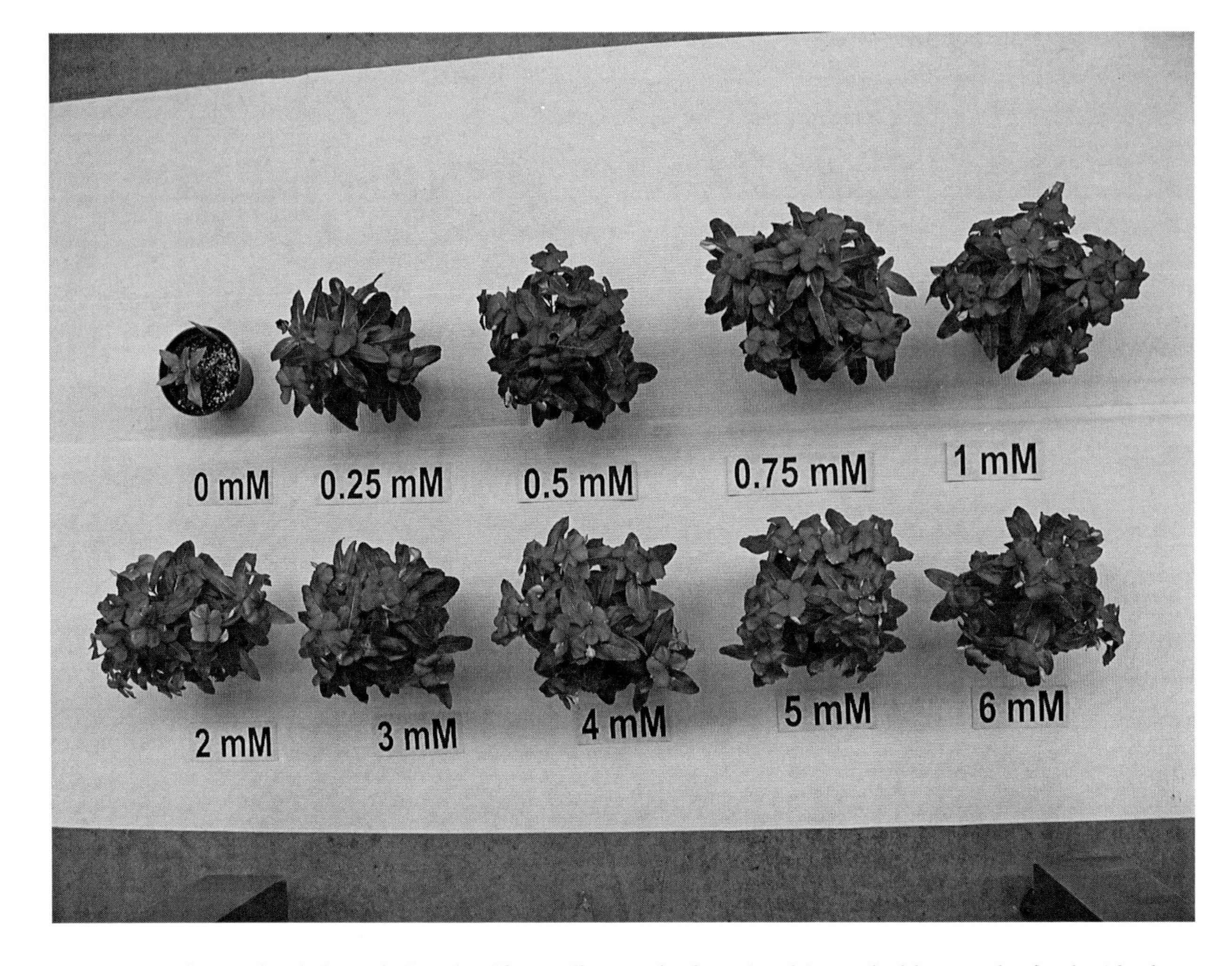

Figure 4-2. The overhead shot of Vinca 'Pacifica Red' research plants is a bit crooked but can be fixed with photo editing software.

hungry, thirsty, or tired. Try to get them to their photo shoot location with enough time to relax but not so much time that they get nervous and begin to pace or try to escape during the wait.

For plant photos, you will need a bright backdrop such as a white sheet or white butcher paper. Some schools have drop-down screens for showing DVDs or for "smart classroom" use (a smart classroom typically has live access to the www. available during lecture as a teaching tool). If you want to use the screen as a backdrop, bring a white plastic trash bag and butcher paper or a sheet to place on the screen, you do not want to leave a water stain on the screen! Black is a newer background color that has been used in horticulture research recently since it allows a lot of contrast with bright green foliage and light colored flowers. Try both background colors to see what looks best against your plants.

Using blocks and boards to set up your photo shoot, you can stack your plants in rows to make a greater impact showing the effects of increased rates of your treatment (Figure 4-3). It takes some time to get the materials beforehand so keep this in mind as you do your planning. When you take the final shot and edit it, you will not even notice the boards (especially if they match your backdrop in color) and can crop out the blocks (Figure 4-4).

Some researchers use photo editing software to cut and paste parts of pictures together to get the image with all treatment representatives in similar fashion to Figure 4-4. This is a wise suggestion for livestock photos since you would not be expected to stack animals on platforms and expect them to stand still for a photo! Animals usually need to be haltered and held by a handler (this is

Figure 4-3. Setting up plants for photo shoot before terminal harvest includes having a good backdrop and way to separate treatments, such as the board and blocks.

true for cattle, sheep, and horses) or led to a feed bunk to feed to reduce movement (hogs and chickens). Livestock experiments are good candidates for captions added directly to the picture in a text box.

Like plants, make sure animals are kept in a cool, shady location until their photo shoot, and use natural lighting if possible. Since most publications allow color figures in addition to grey, set your camera to its highest setting and pixel size for the best quality pictures. Avoid dark, grainy pictures that might mask any fine detail that you are trying to show your readers.

The final harvest side view in Figure 4-4 minimizes the impact of the board and blocks (which have been cropped out of the picture) seen in Figure 4-3 and emphasizes the treatment differences. Note

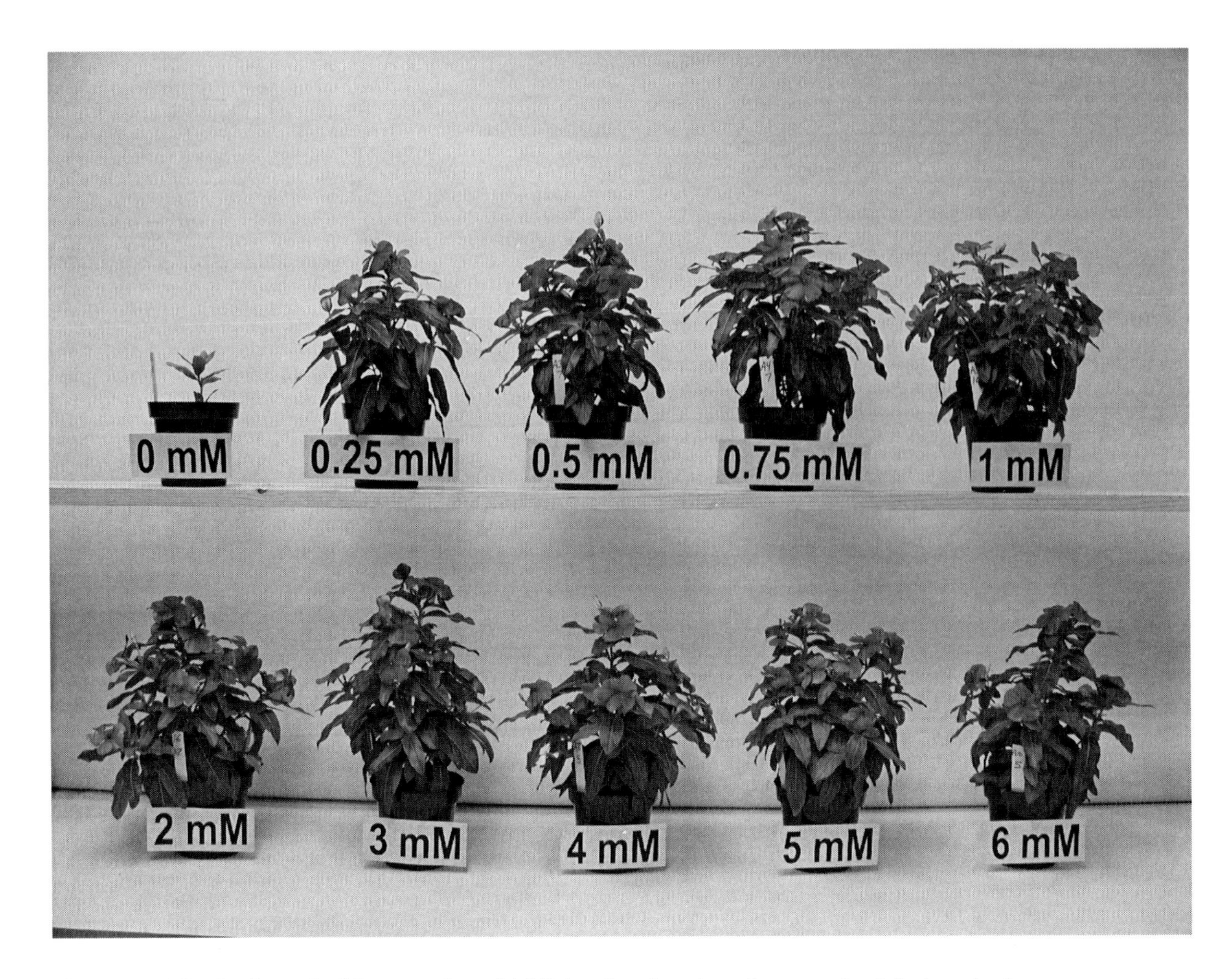

Figure 4-4. The final terminal harvest photo highlights the plants' attributes and minimizes shadows.

how clear and large the labels in Figure 4-4 are, even in a picture that is taken from a distance. Large, easy-to-read labels catch the reader's eye and allow him/her to compare treatments quickly.

Use whatever techniques make you feel the most comfortable and confident but always remember to download your images and check them for clarity as soon as you take them. Enlarge the images on your computer to the size you would use them in a publication or presentation, and always remember to back up your images as well as your written work on another hard drive; they cannot be replaced once your plants are harvested!

For labeling treatments, you can attach index cards with an abbreviation of each treatment using tape to the outside of the pot or by adding the text of the label to the image itself. Avoid handwriting labels, always use a printer or have them printed by the software. The font size and type will need to be adjusted depending on the number of replicates you have in the figure but start by looking at using something simple like Times New Roman, Sans Serif or Arial fonts. Print out samples and see which you prefer.

If you are going to make labels, font size will need to be increased, sometimes up to 100 point, and you will need to determine if bold is necessary. What does a label in 100 point Times New Roman font look like? As you can see in Figure 4-5, this label is quite large. It needs to be if it is going to be seen under a large plant in a pot at a distance with other large plants in the picture. In animal science experiments, images of livestock will require larger labels or some experienced photo editors can create labels on the image with editing software. It is critical to confirm that each treatment is identified correctly in the photo before it gets published or presented.

200 ppm N

Figure 4-5. An example of the actual size of a 100 point Times New Roman plant specimen label for a fertilizer trial.

What can you learn from each figure's caption found in publications? Are they inclusive? Do they help make the figure a stand-alone document in itself? What are the formats for captions found in your model journal? Are they complete sentences or fragments? Do they refer to something specific in the figure? Some journals are very detailed, others not so much. Know what your instructor expects in the caption because all figures (and tables too) must be identified and described with a caption that includes an identifying number. In Figure 4-5 above, the identifying number (the "callout number") is 4-5 and it refers to it being the fifth figure in the fourth chapter of this book.

While you are developing your agriscience experiment that involves living plants and animals, keep in mind that the harvest/termination date is not going to be set in stone. A change in the weather pattern and your plants' growth may be enhanced or delayed. Animals may get sick or may not find the feed in a feed study palatable and therefore will not eat and gain weight and be as big as you expected. Be prepared for delays but know some experiments actually go faster than expected; this is especially true in late spring and summer.

Weekly Data Collection and Creating Data Sheets

Before the experiment is started, you will need to confirm your experimental units are homogeneous, that they are all equal in size, weight, age, or height. If you want to be able to compare "apples to apples," you need to have the initial data checked statistically. We will discuss how to do that more in detail in Chapter 6 but for now you will need to document the initial dependent variables that you want to test later on. Taking initial data is similar to taking pretest data in survey studies and the final data are similar to post-test data. They are compared to see if there has been any change after the treatment is completed. These are standard dependent variables found in experimental design and the variable may be anything you want to measure (e.g., height, average daily gain, dry shoot weight).

If you were studying how different types of fertilizer affect foliage plants and were looking to see which grew the tallest from a fertilizer trial, you would need to measure and record their initial heights. What would a data sheet for this experiment look like? That is an important question because you need to have a document and location on your computer to keep data secure throughout the duration of the experiment. Again, it is recommended to have a hard-bound college composition book and to write down everything, even if you think it is not important at the time. Many pieces of information have been lost in research because someone wrote them down on loose-leaf binder paper, on note cards, or on sticky notes. All of these are easily lost or misplaced but if you write in a composition book that you confirm can never have pages removed, the chances are greatly improved of keeping better records.

Another benefit of the composition book is that you have an instant diary of your tasks you performed for your experiment, a list of important dates, references, or contacts. This can be invaluable for developing additional studies or to review notes when writing up your discussion section of your lab report later on. Again, do not forget to back up your data by entering them into your computer and backing that up as well!

You can print out a data sheet to tape or staple into your composition book (a form you create from a spreadsheet software program such as Excel) to record your data on in the field. Excel is a well-known spreadsheet software package from the Microsoft® Corporation and is a component of Microsoft Office®.

Are there any data sheets that are recommended over others? Not really, but you can get ideas from published lab manuals or you can design your own, changing the fields in the spreadsheet to reflect the data you need to collect. Benefits of using Excel spreadsheet software include its relative ease of use, its availability since most computers have it installed already, and the ability to export files into other software programs for analysis.

To make a data sheet in Excel, you can create columns like those found in Table 3-1 from Chapter 3. Do not reinvent the wheel, if using Excel helps you get better organized for one part of the experiment such as randomizing treatments, use it again for another part such as data collection. Excel is a great tool and can be used to analyze data or be exported into another statistical software product. For our fertilizer example, we want to measure the heights of 30 weeping fig trees (*Ficus benjamina*) grown under 2 fertilizer treatments, one with a time-release fertilizer (Osmocote®, for example) and the other using a liquid fertilizer (using Miracle-Gro®, for another example). You would take half of your foliage plants and assign them to the time-fertilizer treatment and the other half of the plants to the other treatment, the liquid fertilizer, having 15 replicates in each treatment. This experiment would not have a control group since plants need fertilizer and we want to see which fertilizer brand performs best and/or costs less if they perform equally.

For students unfamiliar with Excel, it is simply a spreadsheet program that has the capability of various mathematical and statistical functions with amazing ability to organize data by using rows and columns (Figure 4-6). Note that Figure 4-6 shows only the first half of the second treatment's replicates but you can see how easy it is to make columns for recording each week's data.

The data can be whatever you want to measure: height, dry shoot weight, yield, pounds of milk, average daily gain, food eye appeal and taste, etc. The dependent variable is whatever you think the research will determine best and is limited only by your imagination. In this case, it would be plant height recorded in centimeters (to the nearest tenths place). You need to keep track of the units of measurement for each of your projects, writing them down in your lab notebook and recording them on computer software.

Figure 4-6 has simple headings at the top of each column representing each week's data and makes it easy to record the data in each column. Remember how we showed you how to label the treatments with coding variables from Table 3-1? Those numeric labels correspond to each replicate on this data sheet on the two left-hand columns.

If you were in a greenhouse taking data on fig trees for example, you would select a plant, take its measurement, and record that measurement in the appropriate cell in the appropriate column. That is it; you are done with that plant for the week. Not all experiments require weekly data collection so discuss this with your instructor or research advisor prior to developing your data sheet but remember that it is better to consider taking more data than you need in the beginning (i.e., different types) than to wish you had at the end of the experiment.

Each week you should measure the plants' height (or other dependent variable) in metric units (centimeters for this example) always to the tenths place (or hundredths place if you have the technology in plant, fungi, and bacterial studies), as this will improve accuracy. Livestock weights are

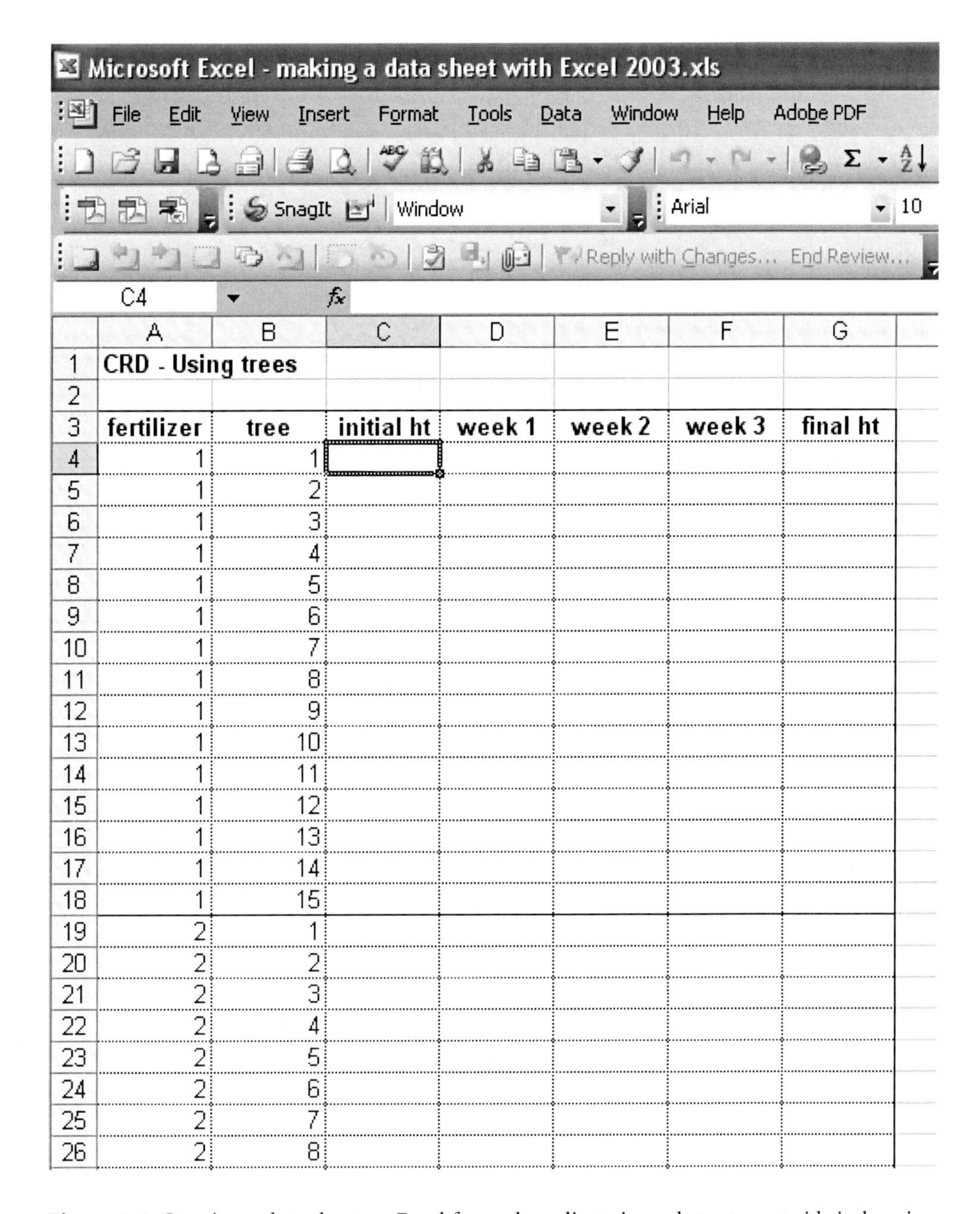

	A	B	C	D	E	F	G
1	CRD - Using trees						
2							
3	fertilizer	tree	initial ht	week 1	week 2	week 3	final ht
4	1	1					
5	1	2					
6	1	3					
7	1	4					
8	1	5					
9	1	6					
10	1	7					
11	1	8					
12	1	9					
13	1	10					
14	1	11					
15	1	12					
16	1	13					
17	1	14					
18	1	15					
19	2	1					
20	2	2					
21	2	3					
22	2	4					
23	2	5					
24	2	6					
25	2	7					
26	2	8					

Figure 4-6. Creating a data sheet on Excel for each replicate in each treatment aids in keeping track of records throughout the duration of the study.

taken in whole numbers. Do not forget to discuss precision with your lab partners: where will you take the measurements on the plants each week? Again, this will lead to improved accuracy and better, more credible results.

In this example there are 30 weeping fig trees undergoing 2 fertilizer treatments and you can enter them in columns by week. Having a hard copy of this sheet in a composition lab notebook makes it easy to use when collecting data in the greenhouse, field, or feedlot where space may be limited and there are probably no smooth or dry tables to write upon.

As you are looking at the growth parameters you chose, be aware of both random and systematic error with measurements. **Random error** (also called human error) is caused by inaccurate measurements or carelessness and can reduce the chance of showing significant results in the data. **Systematic error** comes from having a piece of equipment not work correctly or something tangible that affects the treatments' values throughout the duration of the experiment, (e.g., a scale that is off by 5 grams, the heater in the greenhouse blows on one side of a bench more than another side) causing skewed results. Both random error and systematic error can be reduced with careful planning and attention to detail and if you work with several lab partners, take time to agree on measuring techniques and procedures to ensure precise and accurate results. By the way, we know there is going to be some sort of error; error is the difference between the true value of the measurement (which we will never know) and the measurement that we take ourselves. We get pretty close, that is sure, but we will never pinpoint that measurement exactly. This error is calculated by statistical software programs and will help us determine if our data are significant. We will pick up this discussion again in Chapter 6.

As you take data each week, your spreadsheet will start to fill up with numbers. Create a sheet that works for you, you may use something similar to this data sheet or may copy a data sheet from an existing lab exercise or can develop your own. Data from each plant (replication) can be included weekly in Excel over time (Figure 4-7) so that you can see how they are growing week-to-week and provide a back up copy to your handwritten notes. Make sure you double-check your entries as you record data into your notebook, it is easy to confuse plants when you write in small spaces on the data

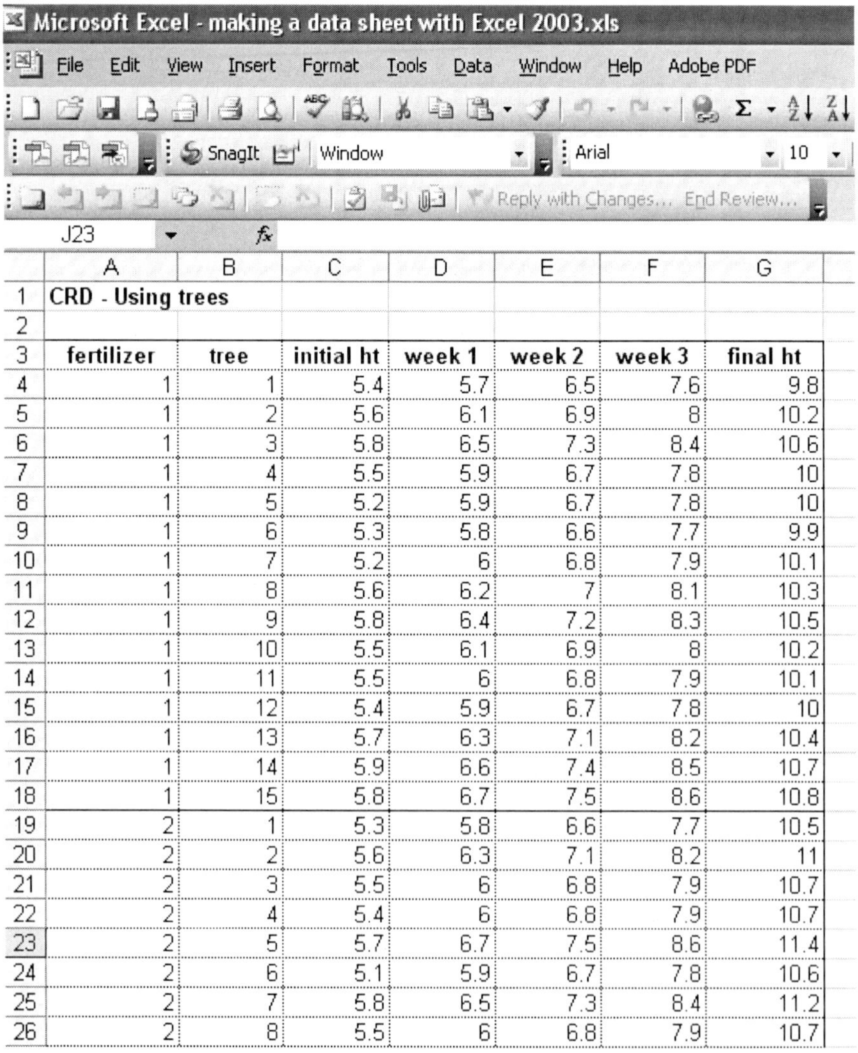

Microsoft Excel - making a data sheet with Excel 2003.xls

File Edit View Insert Format Tools Data Window Help Adobe PDF

SnagIt Window Arial 10

Reply with Changes... End Review...

J23 fx

	A	B	C	D	E	F	G
1	CRD - Using trees						
2							
3	fertilizer	tree	initial ht	week 1	week 2	week 3	final ht
4	1	1	5.4	5.7	6.5	7.6	9.8
5	1	2	5.6	6.1	6.9	8	10.2
6	1	3	5.8	6.5	7.3	8.4	10.6
7	1	4	5.5	5.9	6.7	7.8	10
8	1	5	5.2	5.9	6.7	7.8	10
9	1	6	5.3	5.8	6.6	7.7	9.9
10	1	7	5.2	6	6.8	7.9	10.1
11	1	8	5.6	6.2	7	8.1	10.3
12	1	9	5.8	6.4	7.2	8.3	10.5
13	1	10	5.5	6.1	6.9	8	10.2
14	1	11	5.5	6	6.8	7.9	10.1
15	1	12	5.4	5.9	6.7	7.8	10
16	1	13	5.7	6.3	7.1	8.2	10.4
17	1	14	5.9	6.6	7.4	8.5	10.7
18	1	15	5.8	6.7	7.5	8.6	10.8
19	2	1	5.3	5.8	6.6	7.7	10.5
20	2	2	5.6	6.3	7.1	8.2	11
21	2	3	5.5	6	6.8	7.9	10.7
22	2	4	5.4	6	6.8	7.9	10.7
23	2	5	5.7	6.7	7.5	8.6	11.4
24	2	6	5.1	5.9	6.7	7.8	10.6
25	2	7	5.8	6.5	7.3	8.4	11.2
26	2	8	5.5	6	6.8	7.9	10.7

Figure 4-7. Weekly data entered on an example data sheet on Excel for each replicate in the fertilizer study.

sheet. It might be easier for you to increase the size of the cells for each entry so you do not have to write small during data collection to begin with. This will make it easier to read your handwriting later on when you enter your field data into Excel in the computer lab or at home.

Data from each plant (replication) can be graphed over time to see how certain plants are growing. Also, the average data from each week can be computed by Excel at the bottom of each column and you can use it to compare to other weeks' mean heights in a graph (Figure 4-8). Graphs are often included in publications and presentations since they show data quickly in an easy-to-read format and highlight trends in averaged data. We use means (averages) in scientific research for several reasons: looking at raw data can be overwhelming or confusing, outliers (far out measurements that do not seem consistent with other data) can take attention away from the focus of the dependent variable (the mean), and obvious trends in data can be noticed almost immediately.

If you want to compute means by week, Excel can calculate them for you and you can place the mean in any cell you want. As seen in Figure 4-8, you will need to go to the Insert menu at the top of the spreadsheet and scroll down to **Function.** When you click on the Function tab Excel gives you several options to select the calculation you want it to perform and in this case, we want to compute a

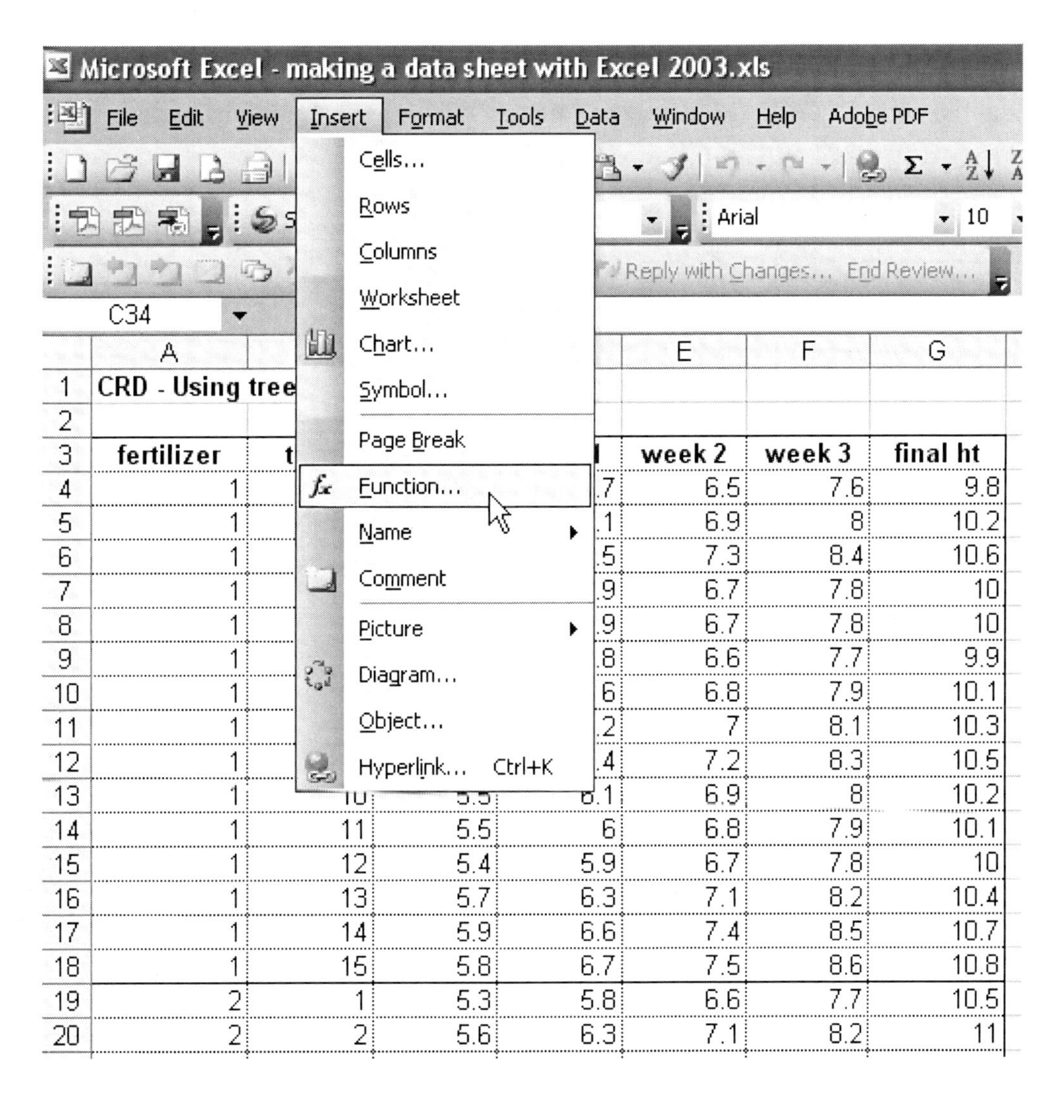

Figure 4-8. To calculate a mean in Excel, you need to go to the Insert menu and scroll down to the Function tab and click on it.

mean (called an **AVERAGE,** as defined by Excel) as shown in Figure 4-9. The definition of each calculation (or function) that you want to make can be found in this dialog box and you can scroll down the list of functions to see what Excel offers to its users. Here is a hint, if you want to specify a cell for your mean to appear in, click on a cell at the bottom of a column, such as C34 in Figure 4-9, and then continue through the steps shown here. When you finish, the mean will be placed in that cell, which in this case was cell C34.

Once you have clicked on AVERAGE and then OK, another dialog box will appear (Figure 4-10) requesting the range of data you want to take the average from. Sometimes you have to delete what first appears in the range and then go highlight the data you want to take a mean from. A word of caution: make sure you are only averaging the data from the replicates in one treatment though, not the entire column. Note that in Figure 4-10 only the first 15 replicates from the first treatment are highlighted and entered into the range in the dialog box. After clicking OK, the mean will be calculated and if you specified a particular cell, it will appear in that cell. In Figure 4-11 the mean of 5.546667 appears in cell C34, just as it was directed to.

While you look at the screen of data in Figure 4-11, note that the function or equation to calculate the mean from cell C34 is found in the white **function bar** (*fx*) near the top of the screen. In the future you can calculate means of any data by clicking on an empty cell where you want the mean to appear, typing in =AVERAGE() up in the function bar, clicking your cursor between the parentheses symbols [()] in the function bar, and then highlighting the data you want to average with your cursor, finishing by clicking the enter key. The mean will be calculated from your highlighted data and will appear in the cell you designated.

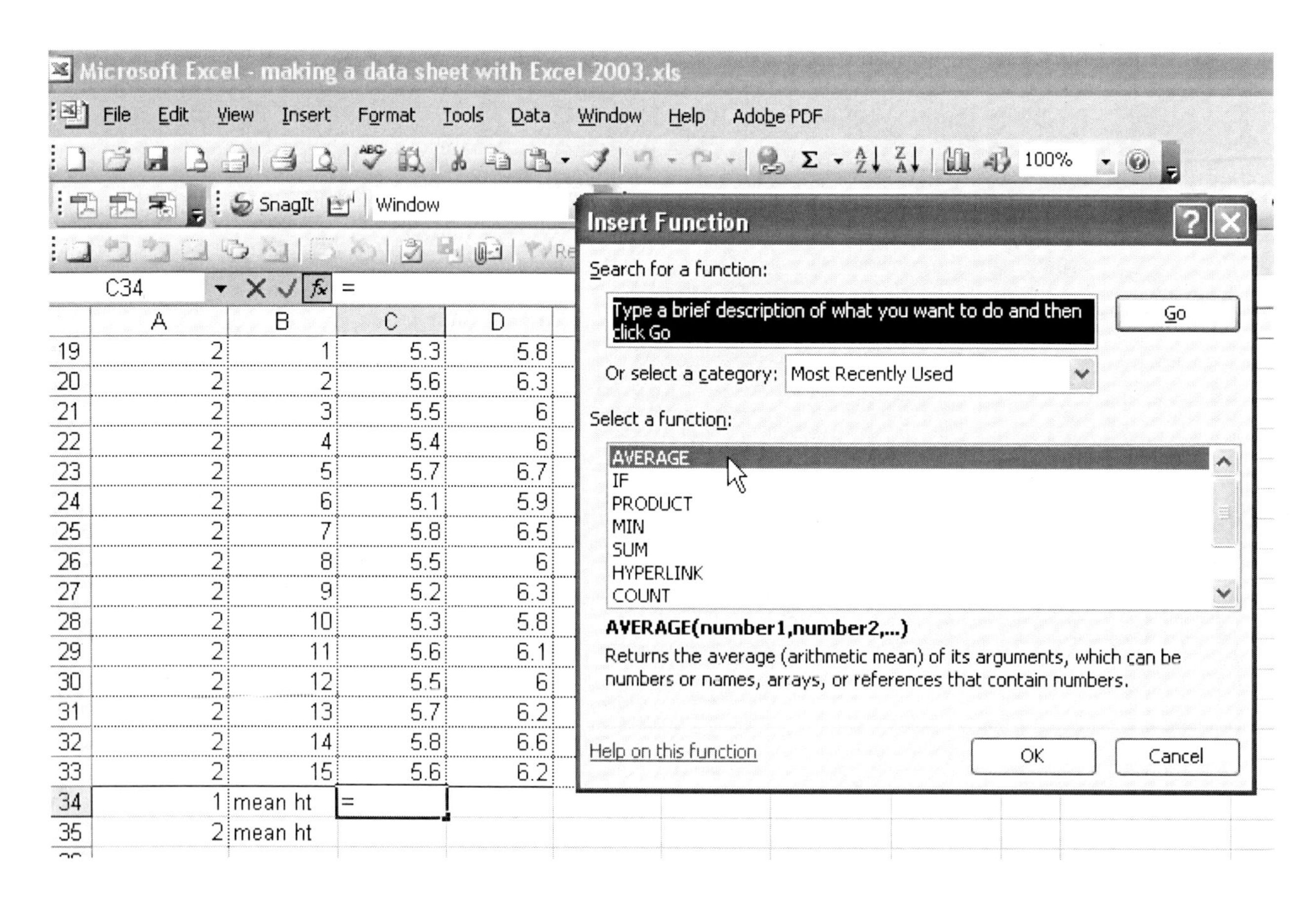

Figure 4-9. The Insert Function dialog box appears and asks you to input the type of computation you want Excel to perform. In this case, you will click on AVERAGE and then OK.

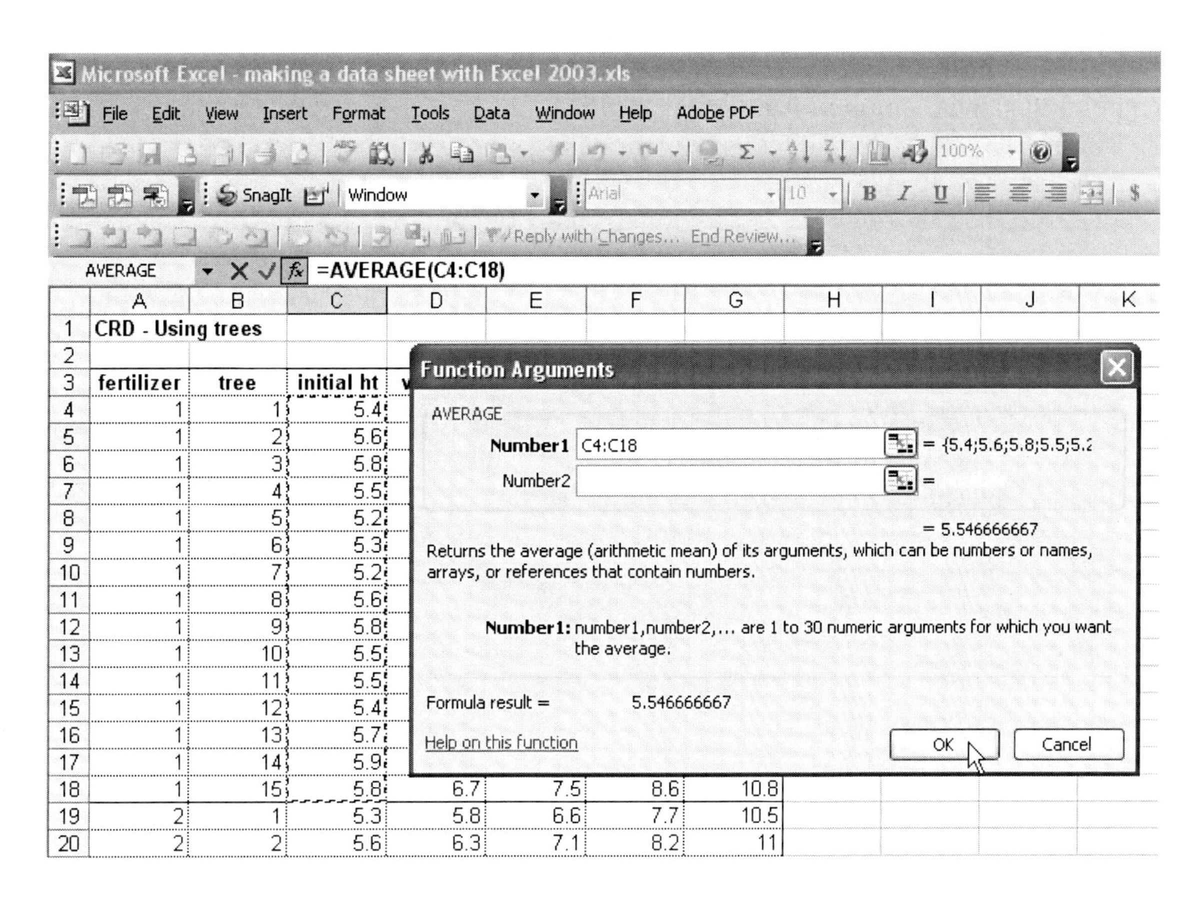

Figure 4-10. The Function Arguments dialog box appears and asks you to input the range of data you want to compute the average from.

Microsoft Excel - making a data sheet with Excel 2003.xls

C34 =AVERAGE(C4:C18)

	A	B	C	D	E	F	G
20	2	2	5.6	6.3	7.1	8.2	11
21	2	3	5.5	6	6.8	7.9	10.7
22	2	4	5.4	6	6.8	7.9	10.7
23	2	5	5.7	6.7	7.5	8.6	11.4
24	2	6	5.1	5.9	6.7	7.8	10.6
25	2	7	5.8	6.5	7.3	8.4	11.2
26	2	8	5.5	6	6.8	7.9	10.7
27	2	9	5.2	6.3	7.1	8.2	11
28	2	10	5.3	5.8	6.6	7.7	10.5
29	2	11	5.6	6.1	6.9	8	10.8
30	2	12	5.5	6	6.8	7.9	10.7
31	2	13	5.7	6.2	7	8.1	10.9
32	2	14	5.8	6.6	7.4	8.5	11.3
33	2	15	5.6	6.2	7	8.1	10.9
34	1	mean ht	5.546667				
35	2	mean ht					

Figure 4-11. The mean of the initial heights from the first treatment appears at the bottom of the first column.

The mean tree heights from our fertilizer example (cells C34 and C35) have several numbers past the decimal point, making them messy and long (Figure 4-12). If you were to right mouse click on the mean height in cell C34, you would see a menu providing several options. Click on **Format Cells**, and then click on Number tab and Number category (Figure 4-12). Reduce the number of decimal places down to one, and then click on OK. In addition to changing numbers' decimal places, other numerical items can be changed to different formats: currency, dates, time, and percentages. Other tabs across the top allow the user to change the data in the cells to different fonts, sizes, styles, and colors, and the cell's border/shading. The dotted inside lines and outside solid line border found on our example data in Figure 4-12 were formatted with the **Border** tab within the Format Cells dialog box.

Once you have your means calculated for the first week, you can highlight one cell and drag it across the row to calculate means for the other columns (Figure 4-13). There will be a small black square symbol (not an outlined white plus symbol) when it is able to be copied or dragged across several cells. You can check to verify the means in the "dragged formulas" are mathematically correct by clicking on the computed mean in question, then clicking on the cell up in the function bar and seeing if the data highlighted is the same as the data in the range in the function bar. In Figure 4-13, both treatments' data have been averaged and are located at the bottom of each week's raw data. They seem to be very similar, does this mean we could select either fertilizer to use and get the same results? If so, we would want to know what each fertilizer has in it and what each one costs before making our final decision on which brand to use.

Take some time to play around with Excel's features with some data, changing fonts, sizes, making borders, computing data, and merging cells. The more you practice using Excel, the easier it will become for you.

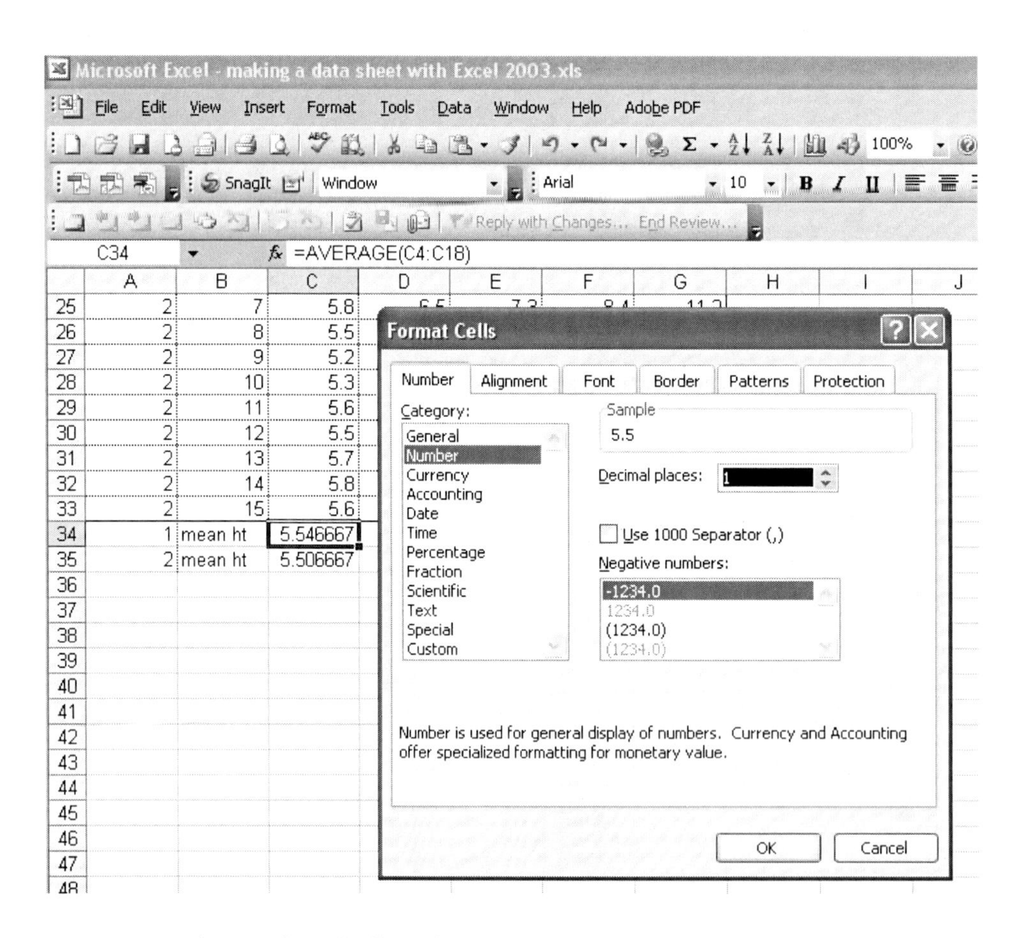

Figure 4-12. The Number tab allows for changing the appearance of numbers with several decimal places or none at all.

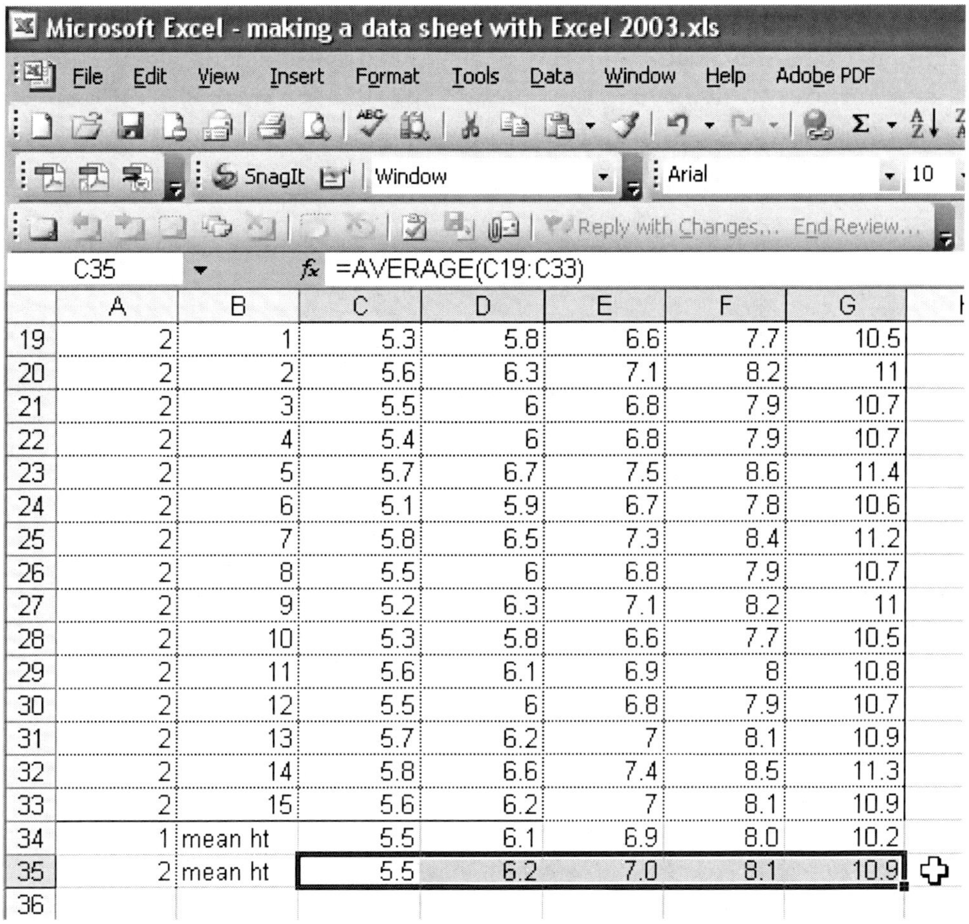

Microsoft Excel - making a data sheet with Excel 2003.xls

C35 =AVERAGE(C19:C33)

	A	B	C	D	E	F	G
19	2	1	5.3	5.8	6.6	7.7	10.5
20	2	2	5.6	6.3	7.1	8.2	11
21	2	3	5.5	6	6.8	7.9	10.7
22	2	4	5.4	6	6.8	7.9	10.7
23	2	5	5.7	6.7	7.5	8.6	11.4
24	2	6	5.1	5.9	6.7	7.8	10.6
25	2	7	5.8	6.5	7.3	8.4	11.2
26	2	8	5.5	6	6.8	7.9	10.7
27	2	9	5.2	6.3	7.1	8.2	11
28	2	10	5.3	5.8	6.6	7.7	10.5
29	2	11	5.6	6.1	6.9	8	10.8
30	2	12	5.5	6	6.8	7.9	10.7
31	2	13	5.7	6.2	7	8.1	10.9
32	2	14	5.8	6.6	7.4	8.5	11.3
33	2	15	5.6	6.2	7	8.1	10.9
34	1	mean ht	5.5	6.1	6.9	8.0	10.2
35	2	mean ht	5.5	6.2	7.0	8.1	10.9
36							

Figure 4-13. Dragging a formula to adjacent cells is a fast way to calculate means down each column. The formula is pulled across and corresponds to the data in the columns above it.

Creating Graphs with Excel

A benefit of using Excel is the ability to create graphs and thus being able to compare data and treatment means in a figure other than in a table. Tables of numbers are fine to look at but most people who are learning about scientific data tend to get bogged down or intimidated by all of the numbers in the rows or columns. Seasoned researchers know how to "read graphs" quickly to look at data and determine what trends appear in the data, and with a little practice, you will be able to do that too.

As part of your lab report, you will need to include a graph and caption that you created from your data. Handwritten graphs are unacceptable so you will need to learn how to make one on a spreadsheet or graphing software program. Excel includes a Chart Wizard menu that you can access with a button (Figure 4-14) that saves you time as it guides you through several graph-creating steps. The button might be placed anywhere on the toolbar so take a moment to find it. There are many types of graphs/charts: column, bar, line, pie, area, and scatter. For clarity, the XY Scatter graph is preferred in this course. Later on you can add a trendline to connect the data points in this type of figure (we will show you how) but initially let us keep it simple by only creating a scatter graph from only one fertilizer treatment with data points from the means of that treatment.

Take a moment and see how we highlighted the first row of means from the first treatment at the bottom of Figure 4-15 before clicking on the **Chart Wizard** button. This sets up the Chart Wizard with the data we want already included in the graph (so we do not have to enter it manually later on). Next, the Chart Wizard will take you through step-by-step the procedures necessary to create a graph.

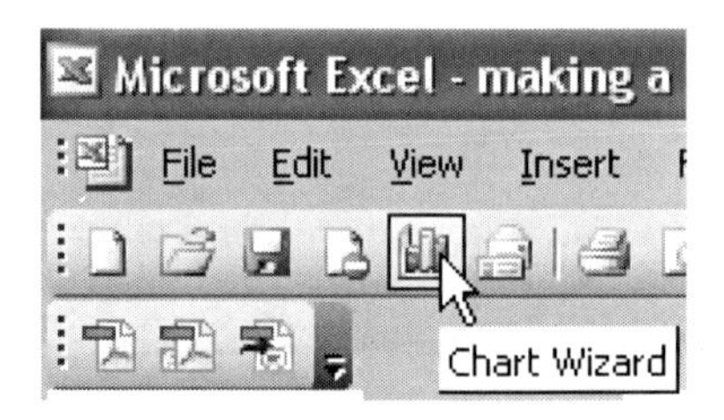

Figure 4-14. Graphing with the Chart Wizard button on Excel begins by clicking on the wizard button.

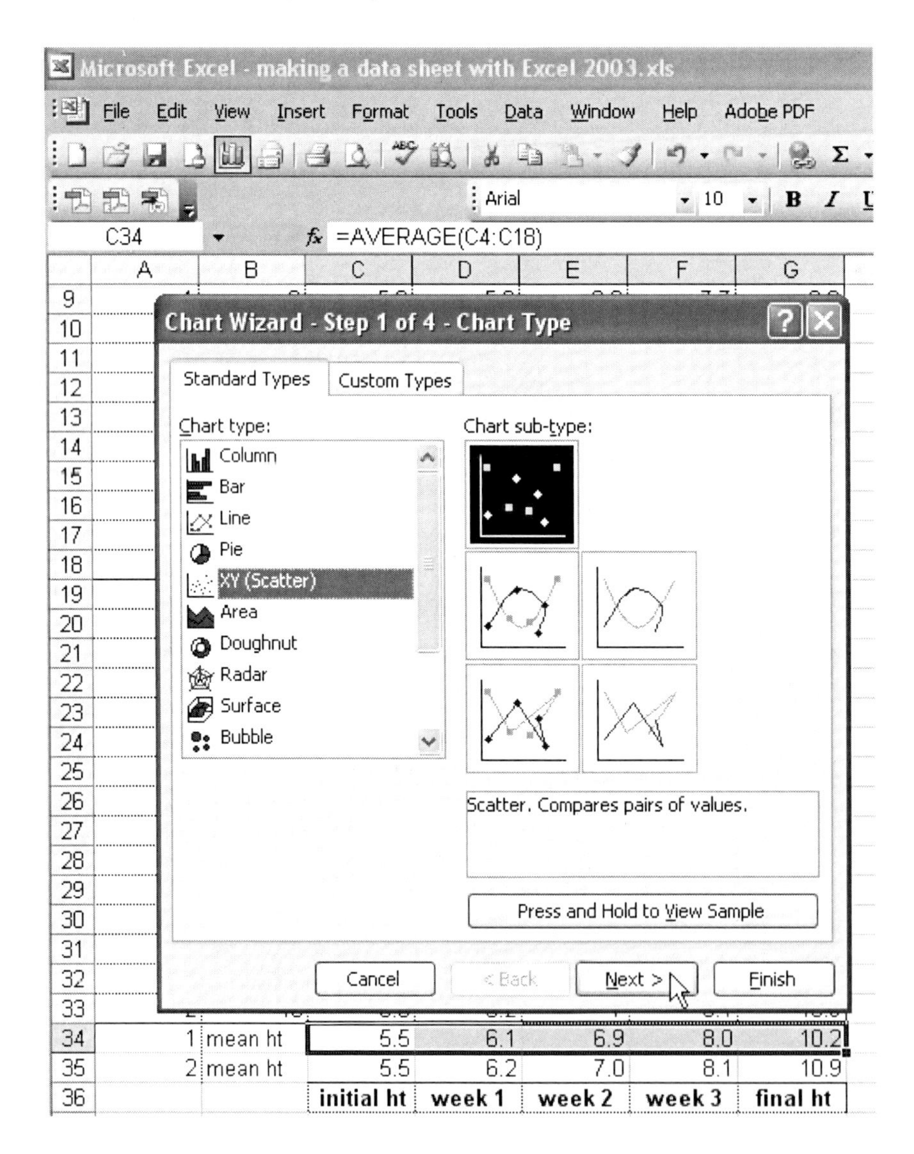

Figure 4-15. Step 1 of making a graph includes selecting the chart type. After highlighting the data you want to include in the graph, click on XY (Scatter) and then click on Next.

In Figure 4-15, you can look down the left-hand column of chart types and find the XY (Scatter). This graph type results in a very clear and easy-to-read figure that can easily be edited for title and axes labels. Note that the Chart sub-type that we want is already highlighted in black (it is the one without any connecting lines). Click on Next to continue to Step 2.

Step 2 of the Chart Wizard (Figure 4-16) shows a rough draft of what your graph will look like in shape and color. Verify that since your original data that you highlighted to include in the graph was in a row that the series of data graphed is also in that row (see how you can select between row and column) in the **Data Range.** Click on Next to go to Step 3 of the Chart Wizard.

Step 3 of the Chart Wizard (Figure 4-17) prompts you to enter a title for your graph and labels for the x and y axes. Do not forget to include the proper units for the y axis (the dependent variable) and give a good description of the x axis (the independent variable). You will see how it looks on the

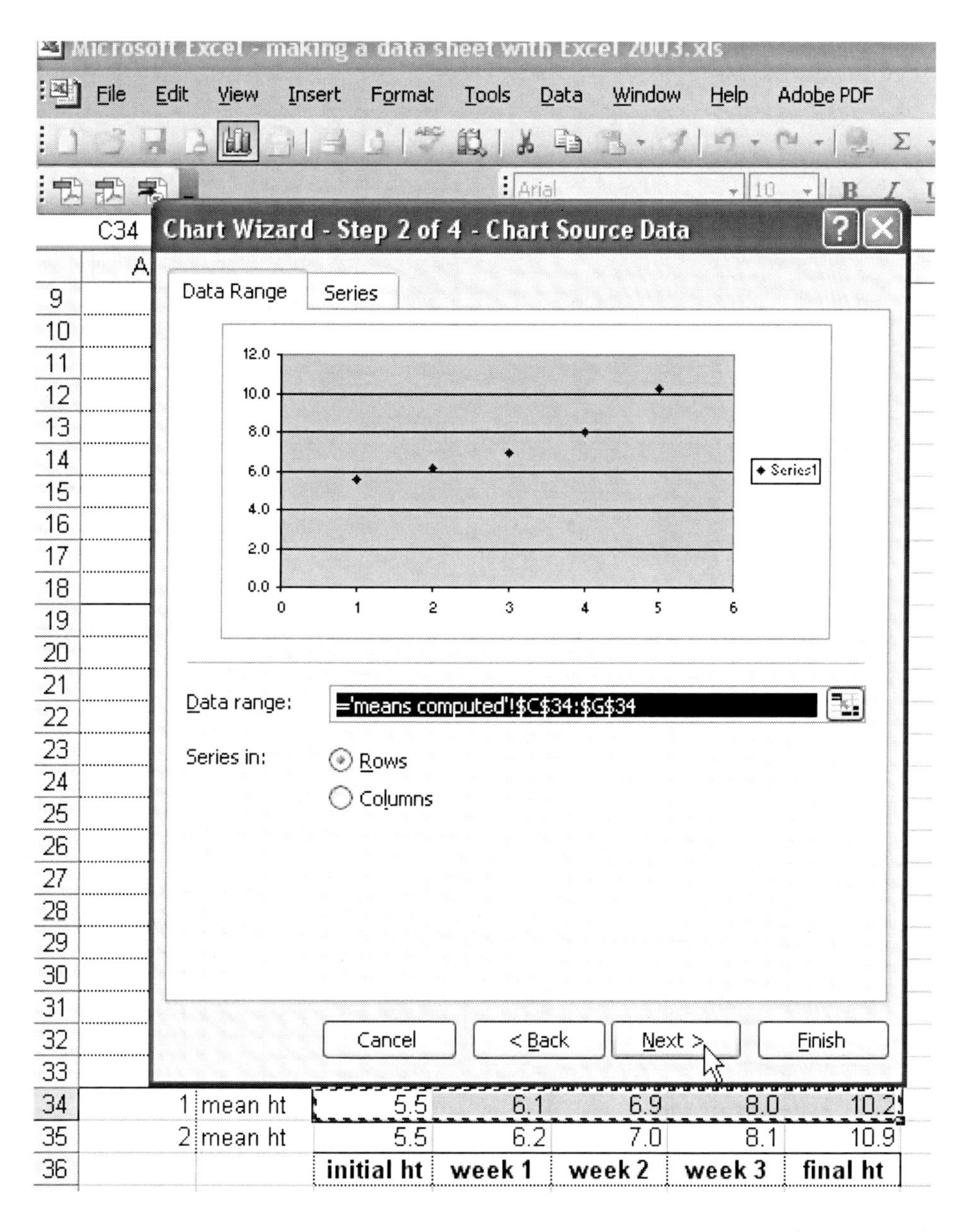

Figure 4-16. Step 2 of making a graph includes selecting the data range. Since the data were already highlighted, an example of the graph will appear.

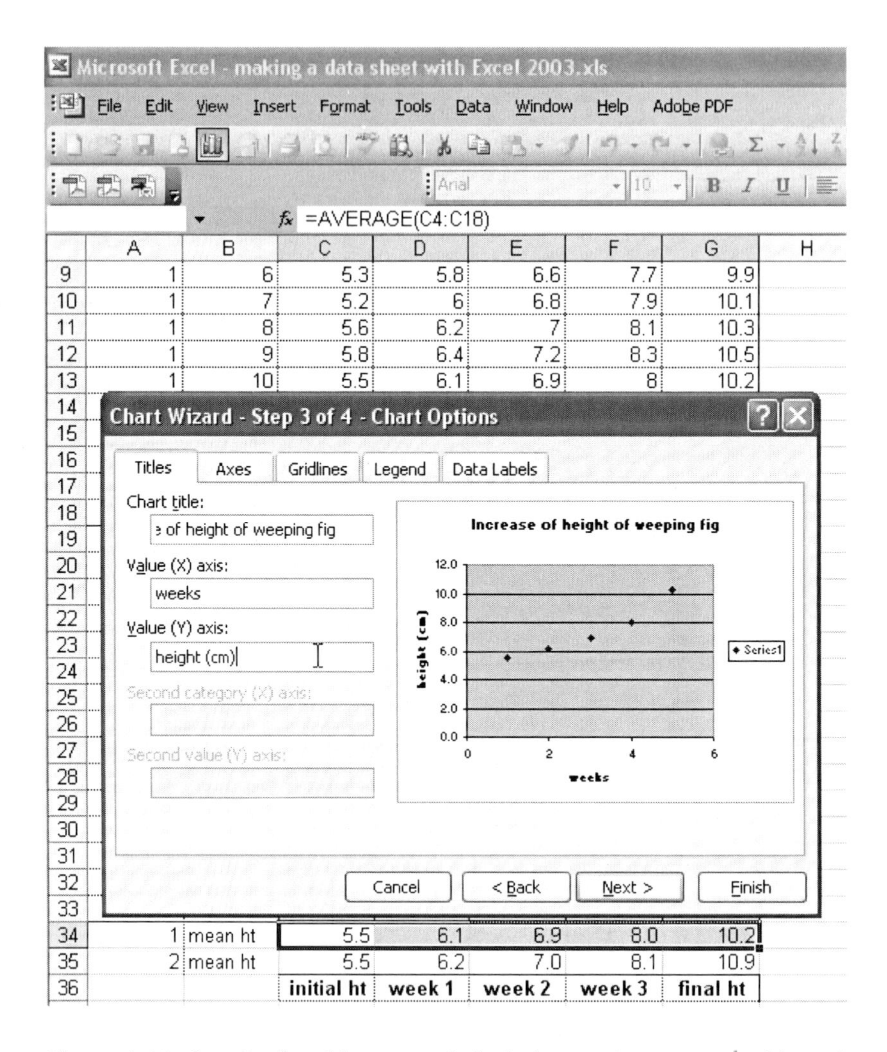

Figure 4-17. Step 3 of making a graph includes creating a graph title and labeling the x and y axes.

rough draft of your graph so make changes if you need to at this point. You can always edit your title and labels later on if you change your mind.

Step 4 of the Chart Wizard asks you where you want the graph to be located; you can either place it on a new sheet in your Excel document or within the current sheet that is already open. The latter is preferred in case you want to make changes; you already have the data for that graph easily accessible (Figure 4-18).

At this point you will have a final version of your graph that you can use in your lab report or in a presentation. You can alter the title (the one in Figure 4-18 is not very descriptive) or change the axes labels to change the overall appearance of the graph just by clicking on them, then correcting misspellings or changing words, font sizes, etc. In case you want to edit your graph to change the background color or remove the lines (Figure 4-18) or legend, you can right mouse click on the lines to clear the gridlines and again on the legend to delete it. You can also right mouse click on the gray background and select a menu called Format Plot Area (Figure 4-19). Change the Border and Area backgrounds to None; a white background should appear on your graph (Figure 4-20).

Removing the background color and lines presents a cleaner, more professional figure (Figure 4-20) that directs the reader's attention to the data points and axes. Look at other figures from journal articles to get an idea of what is expected in your lab report. The graph in Figure 4-20 is ready to be

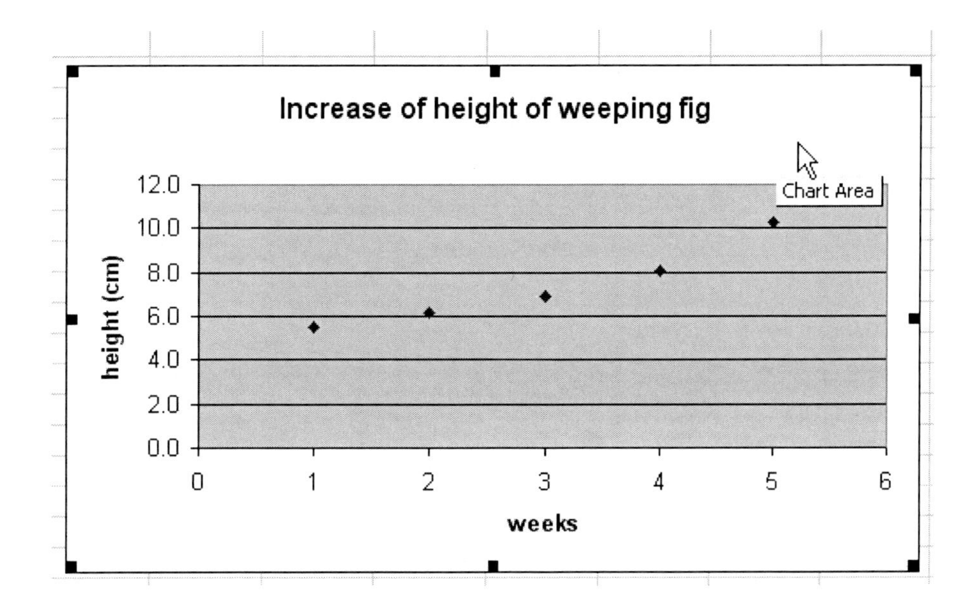

Figure 4-18. Step 4 of creating a graph includes the final graph with appropriate title and labeled axes.

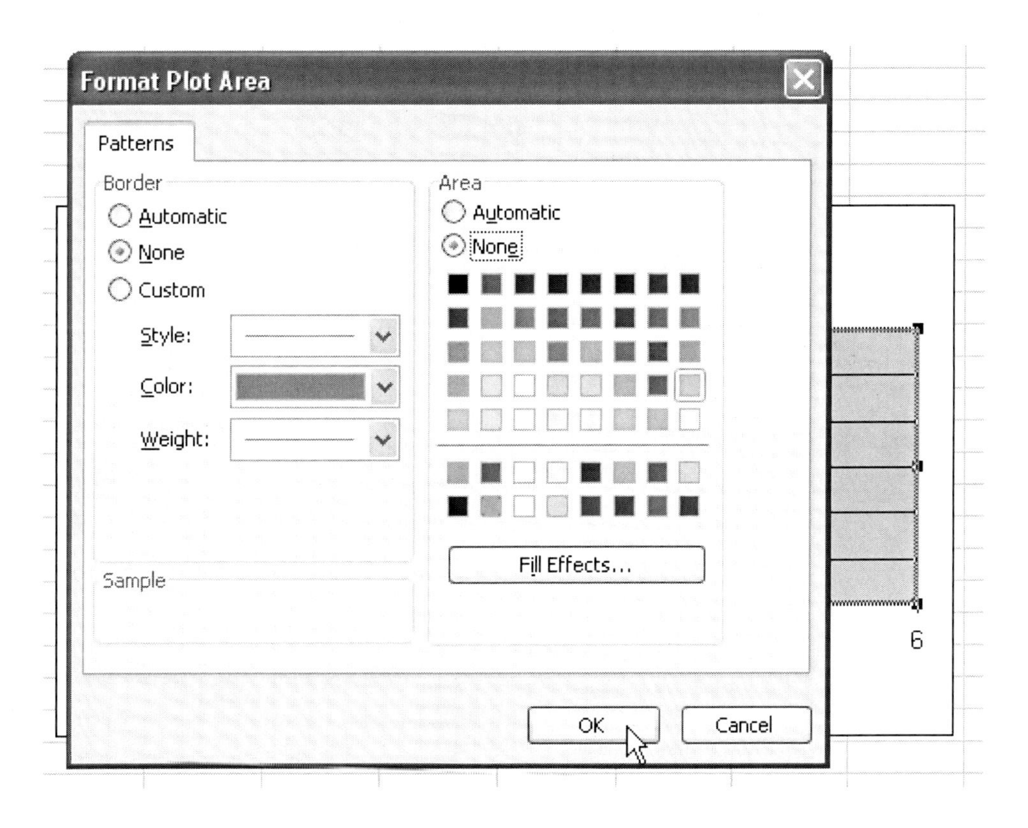

Figure 4-19. An additional step of creating a graph includes cleaning up the background by removing colors and lines from the graph itself. Click on None for each Pattern.

copied and pasted into other documents but will still need to have a descriptive caption written to accompany it.

Suppose you wanted to compare the final heights of the weeping fig trees from both fertilizer treatments. You could make a graph that would show both sets of data, giving a quick overview of how each treatment did. You can follow the steps of using the Chart Wizard to make your graph but you

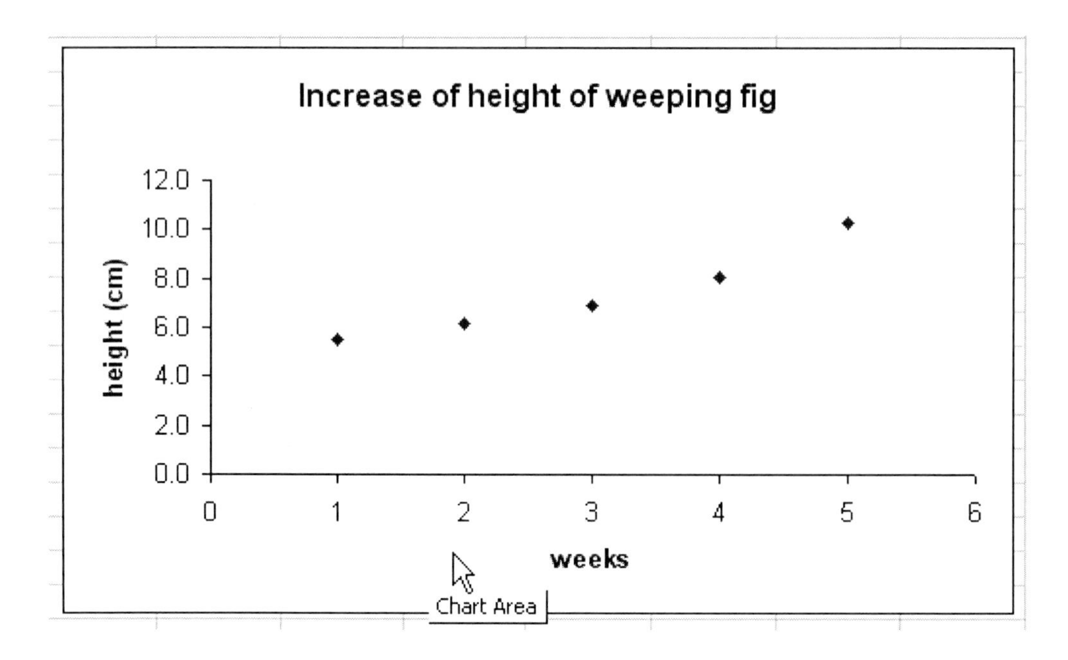

Figure 4-20. A final version of the fertilizer treatment graph appears after cleaning up the background and removing horizontal lines.

will first need to place your data into three columns by copying and pasting from the original data sheet, the first column being the individual plant numbers (Figure 4-21). Figure 4-21 shows that we have added the fertilizer number headings to each column to help identify which data came from either fertilizer in the graph.

Now you can go through the Chart Wizard procedure again using the XY (Scatter) chart to create your graph. Having the two headings (fertilizer 1 and fertilizer 2) will allow Excel to separate the treatments for you, and will place their data into two different colors when you look at them on the computer monitor. In black and white print, the two treatments are distinguished by different symbols (squares and diamonds) for each treatment (Figure 4-22). It is pretty obvious that the plants from the second fertilizer treatment have grown taller than those from the first fertilizer treatment (Figure 4-22). Looking at the data in a graphical form illustrates this easier-to-read format for some readers, when compared to the original columns of data (found to the right of the graph) that can appear clumsy.

What if the data points from both treatments were not so obvious to distinguish? What if they were not split so evenly (half above, half below), and were more mixed up? Is there another way to tell which treatment made greater gains in height? There is, and the feature in Excel is called adding a trendline. A trendline is technically the regression line of the data points in a set of data, and appears near the middle of the data points when added to a scatter graph. It is a good indicator of the data points' direction.

To add trendlines to our data in the scatter graph, you need to right mouse click on any data point from one treatment (Figure 4-23), and click on **Add Trendline.** A dialog box will appear asking what type of trendline to make, click on Linear. For most of our data in this course, we will only need to use linear trendlines, which is a best-fit straight-line equation for simple data sets. To create an additional trendline for the second fertilizer treatment, click on any one of the data points from the bottom set of data (Figure 4-24). Continue with the steps as described earlier.

The other trendline types that were given as options under the Add Trendline menu are used in regression modeling (we will discuss that in Chapter 6) and are reserved for data that change within

G	H	I	J	K
final ht		plant	fertilizer 1	fertilizer 2
9.8		1	9.8	10.5
10.2		2	10.2	11
10.6		3	10.6	10.7
10		4	10	10.7
10		5	10	11.4
9.9		6	9.9	10.6
10.1		7	10.1	11.2
10.3		8	10.3	10.7
10.5		9	10.5	11
10.2		10	10.2	10.5
10.1		11	10.1	10.8
10		12	10	10.7
10.4		13	10.4	10.9
10.7		14	10.7	11.3
10.8		15	10.8	10.9
10.5				

Figure 4-21. Setting up a dual comparison of final height data from two separate fertilizer treatments requires placing data into columns side-by-side.

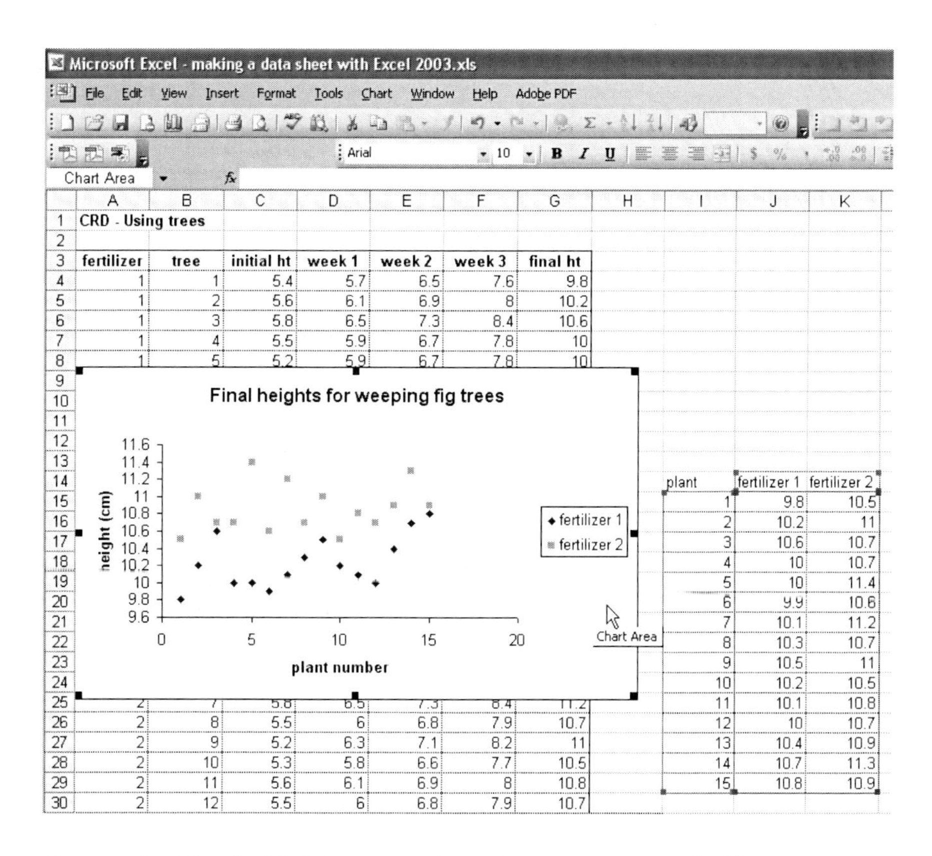

Figure 4-22. The final height comparisons from two separate fertilizer treatments shows a difference, the first treatment having greater height values than the second treatment.

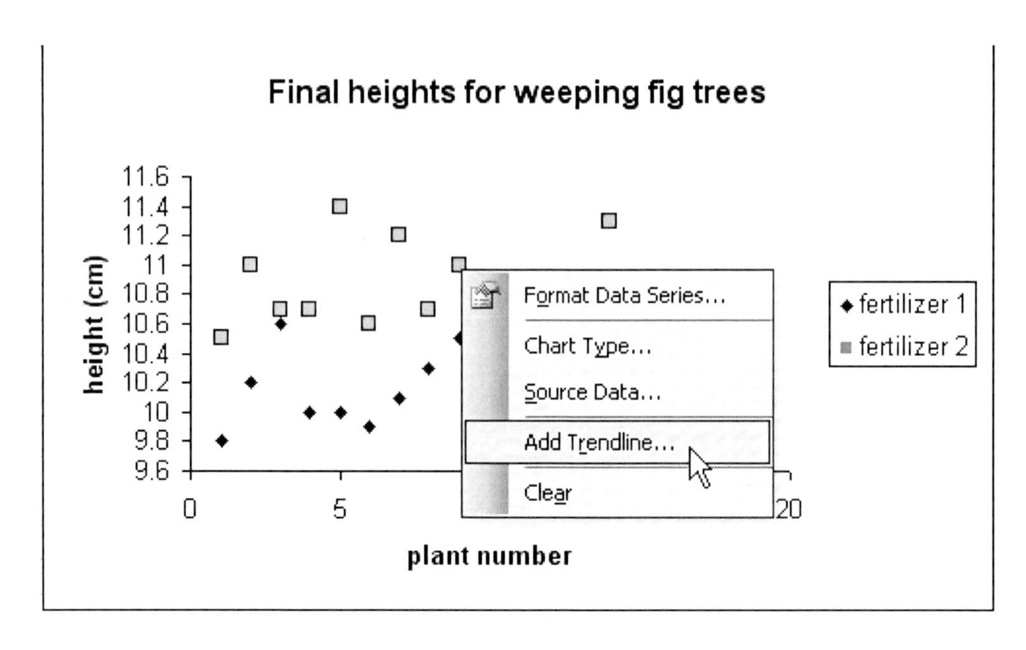

Figure 4-23. Trendlines can be added to graphs to emphasize the mean of those data points.

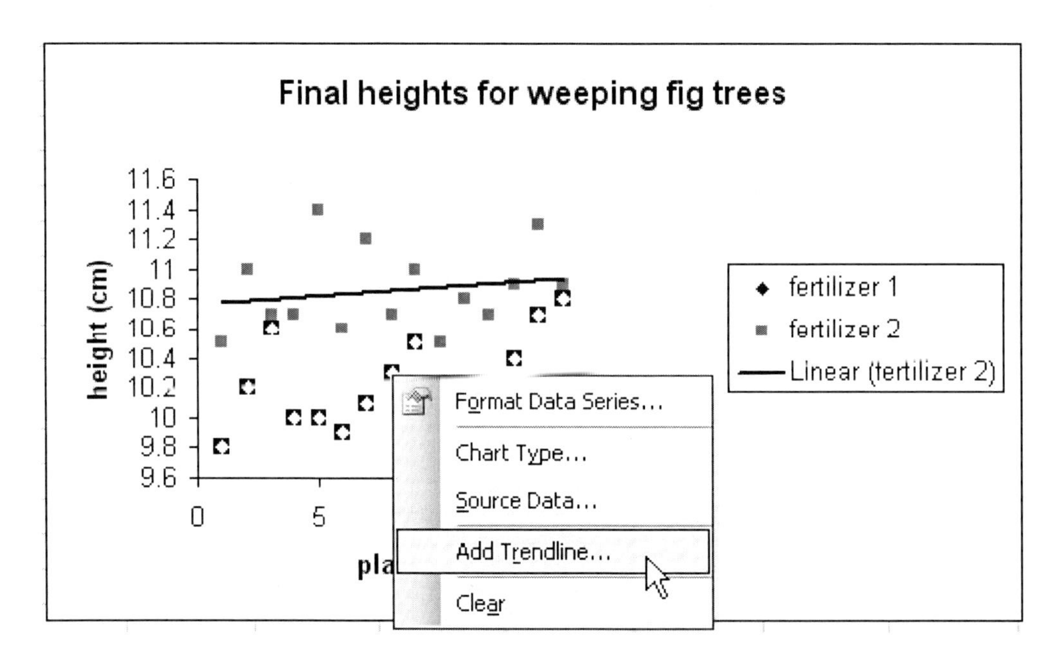

Figure 4-24. A second trendline can be added to graphs to allow for a quick comparison between data sets.

the data set (e.g., rates that increase or decrease). Since our data compare separate plants within different treatments and look at specific, concrete heights, we will use the linear trendline option.

For advanced Excel users, if you are familiar with regression, right mouse clicking on the trendline will bring up a Format Trendline menu. Click on the Options tab to **Display equation on chart** and **Display R-squared value** on chart. The equation for the trendline and the R-squared value will appear near the trendline. We will discuss R-square more in depth in Chapter 6 but know for now that R^2 is the amount of the dependent variable y that is correlated or associated by the independent variable x. High correlations of R^2 can support our alternative hypothesis that there is a difference between treatments. Again, we will discuss more about this later and if you are not familiar with these features, do not worry about them now.

Inserting trendlines only takes a few moments but can save you time when interpreting data, highlighting the differences between data sets quickly. In the example Figure 4-25, we can see the plants from the second fertilizer treatment made greater gains in height than the first treatment, simply by looking at the trendlines.

If you want to format the size of the trendlines to distinguish them from each other, right mouse clicking on one of the trendlines brings up a **Format Trendline** option. You would click on the Patterns tab, then on Custom, changing the Style, Color, or Weight, depending on how you wanted to trendline to look. Finish by clicking on OK. You change the trendline back to its original style by repeating the steps to reformat the trendline.

What if you wanted to make another type of graph, perhaps a bar or column graph or pie chart? For our fertilizer data, a column graph would also be a good figure choice for this example. Using the same data that we used to make the scatter graph to compare both treatments, the only thing you would change when going through the steps of the Chart Wizard would be to highlight the fertilizer 1 and fertilizer 2 columns (columns J and K) and then select Column as a different chart type (Figure 4-26).

Follow the Chart Wizard steps to include a title and x and y axes, and then click on Finish (Figure 4-27). While the graph is being constructed by Excel, you will see the two highlighted columns of data off to the side. You can check these columns that are highlighted to confirm that the graph you are creating is really based on the highlighted data from your fertilizer treatments that you want to compare.

When you have gone through the Chart Wizard steps, the final column bar graph should appear (Figure 4-28), again showing some obvious differences between both fertilizer treatments when compared to the scatter graph from Figure 4-22. We left the gridlines on Figure 4-28 to call attention to the heights of each plant since it might be difficult to compare the levels of the heights across to the units on the y axis. Clearly the dark bars of fertilizer 2 appear taller than the lighter bars of fertilizer 1.

When looking at the data on the column graph (Figure 4-28), remember there are 2 plants at each number on the x axis because there were 15 replicates/treatment, totaling 30 plants in the experiment.

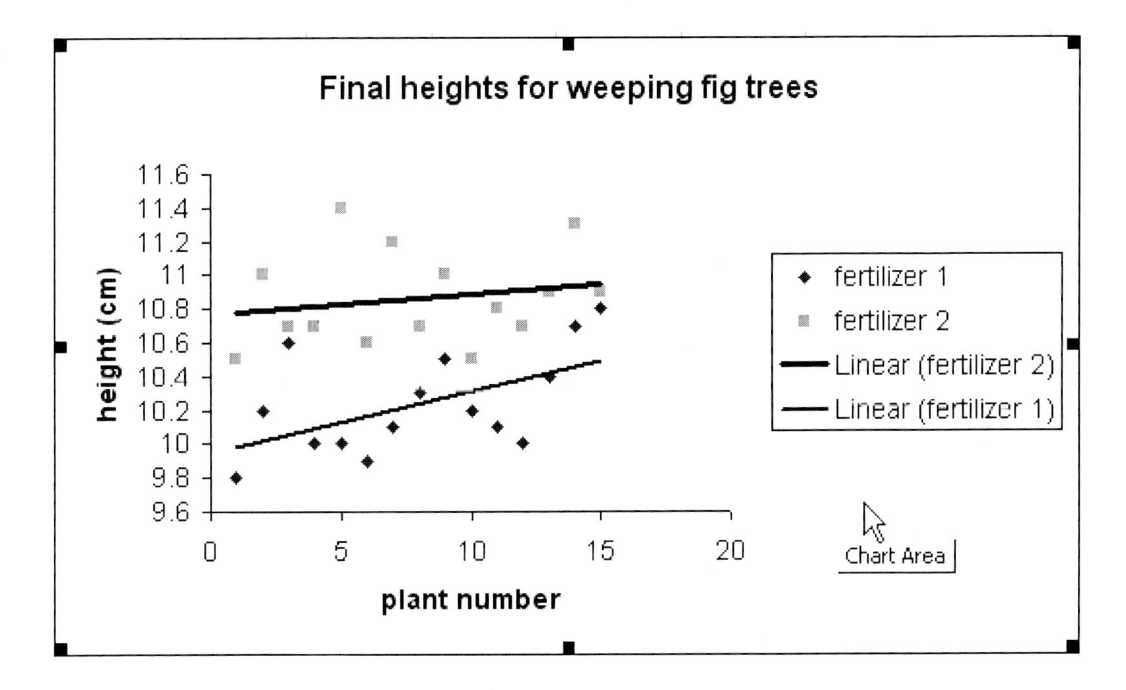

Figure 4-25. Linear trendlines of both data sets help the viewer quickly determine differences between data sets.

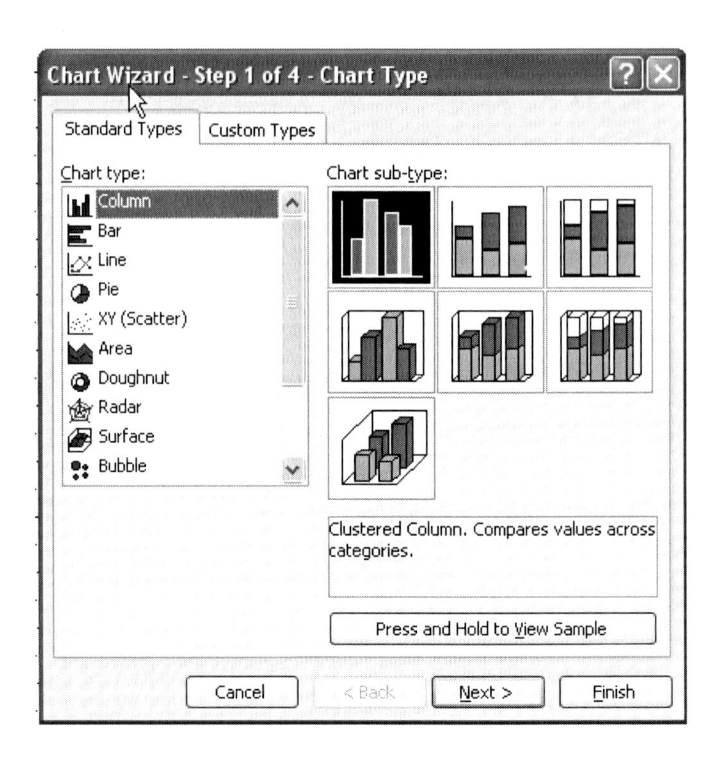

Figure 4-26. Switching between different chart types in Excel is as easy as clicking the mouse. Previews of each chart type are shown on the right of the dialog box.

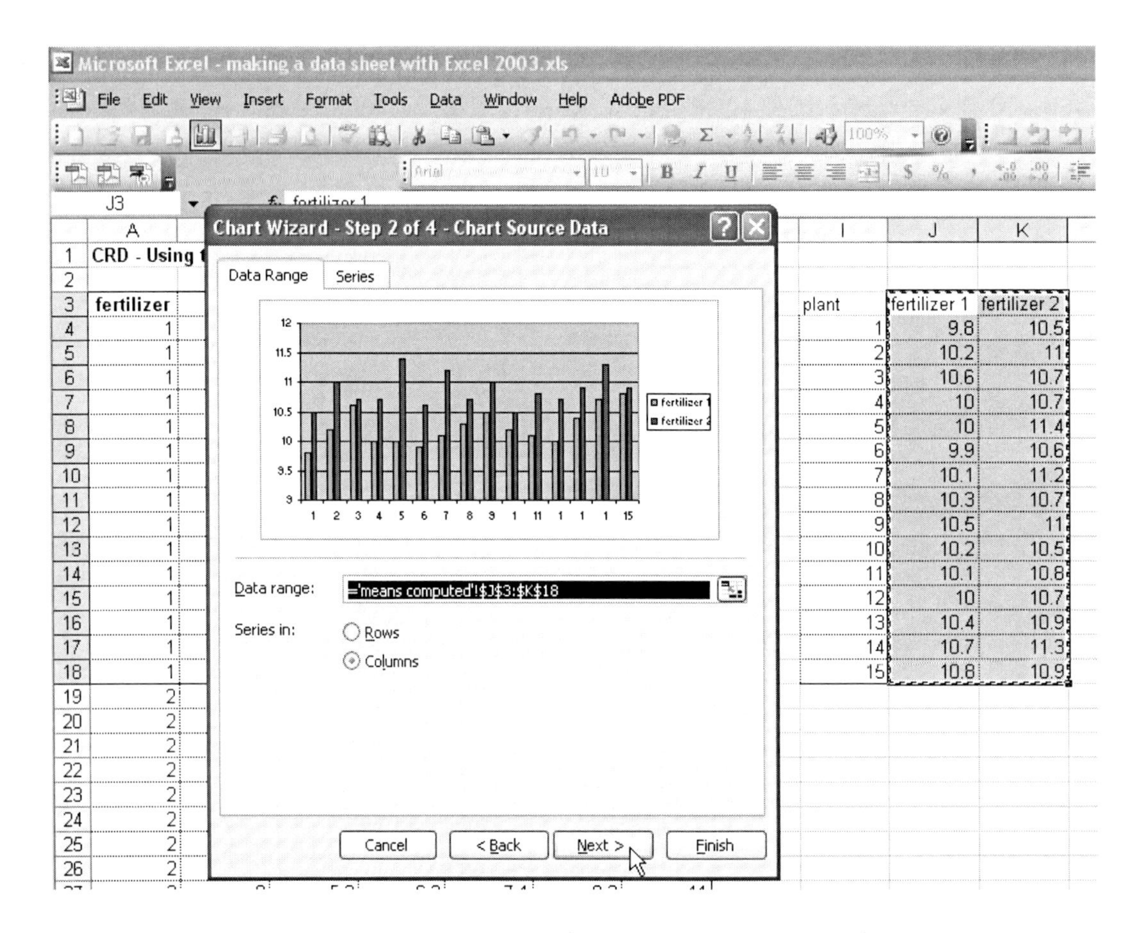

Figure 4-27. The Chart Wizard is useful tool when making different types of graphs for different uses.

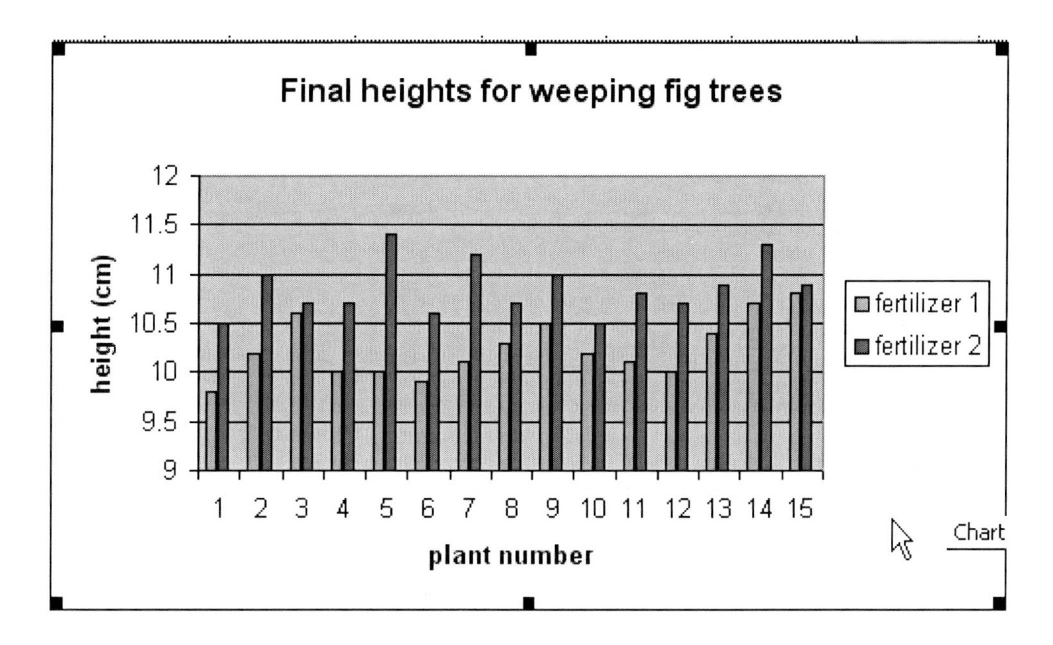

Figure 4-28. The final height comparisons from two separate fertilizer treatments in a column graph form, showing the first treatment having greater height values than the second treatment.

This is similar to what you have seen in Figure 4-22, where there are two data points are each plant number as well, one plant (replicate number 5, for example) from the first treatment and the other replicate (i.e., number 5) from the second treatment.

Do you see any other trends or patterns in the data? There are some outliers, some plants that grew really tall and some that did not grow as much. Since we see that trend within both groups of plants, we can hypothesize that it might be due to the plants themselves. We will discuss what the data might be trying to tell us more in Chapter 6 on data interpretation.

Other data sheets and graphs can be created for your agriscience experiments depending on the numbers of independent variables, numbers of replicates, and dependent variable measured. If you wanted to measure several types of dependent variables, all you need to remember is that you need a column for each one that you want to add (in livestock: weight, average daily gain, wool length, or pounds of milk; and in plants: dry shoot weight, tons of almond meats, tree height, board feet of lumber).

Complete the next exercise using the journal articles selected by your instructor. Please note that you will need to have access to Excel software to create the graph for the final question. All papers need to follow the printing format of the syllabus (black ink, etc.)

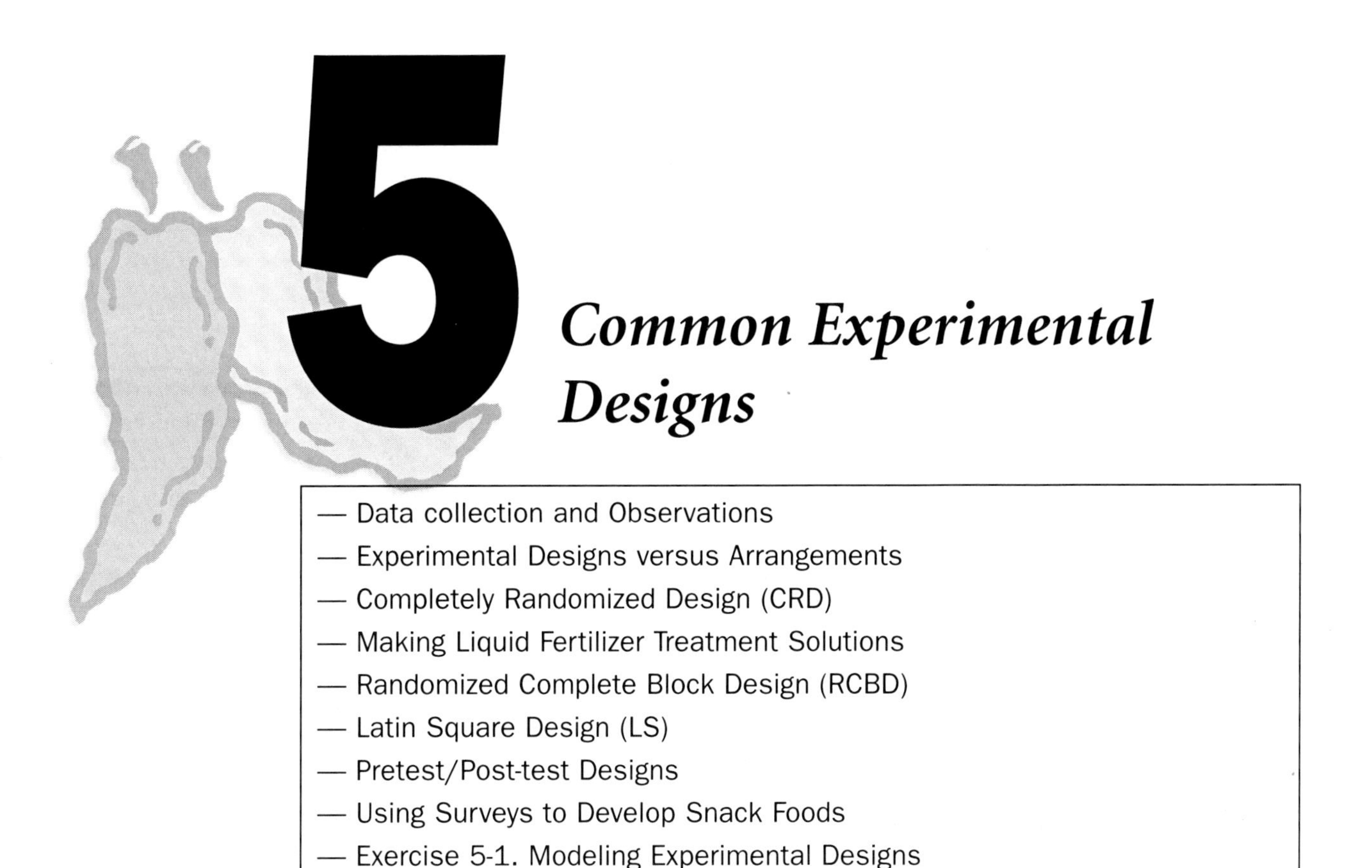

5 Common Experimental Designs

Data Collection and Observations

Once you have your experiment planned, your objectives written down, your data sheets developed, and graphs and other figures created, it is time to determine how you are going to analyze your results. Comparing data is a key concept in science investigations so once you know what dependent variables you are going to compare, you can move on to deciding how you are going to compare them. This will involve some basic statistical analyses or arrangements (which we will cover in more detail in Chapter 6) but do not get upset if you do not know statistics! We will teach you what you need to know to understand the analyses.

There are several common statistical methods we are going to learn about that will help determine if there are statistical differences in our data, and that is why we use them. Since we want to compare the effects of our treatments on a dependent variable (height, weight, yield, etc.), we need to set up our experiments with specific designs in mind to give us that ability. This chapter will cover some common experimental designs and how we can use them to eliminate or reduce as much variation as possible in our experiments. From there, we will see which analyses work best with our data.

Experimental Designs versus Arrangements

Experimental designs are named by how the experimental units are randomized in the field, pasture, feedlot, field, nursery, or on a greenhouse bench. Arrangements are how the treatments will be statistically related to each other. For example, a single independent variable with several levels is a completely randomized design and the statistical arrangement is a 1-way factorial, typically analyzed with analysis of variance (ANOVA). Since you already know how to randomize experimental units into treatments from Chapter 3 to reduce variation and bias, we will go into more detail about how you can eliminate other sources of variation.

The main thing to remember in credible scientific investigations is that we need a way to gather quantitative or qualitative evidence that shows that there may be differences between treatments. How we gather that evidence is enhanced or hampered by the way the experiment was set up in the beginning. It is always better to review your design and analysis with a fine-tooth comb before you start the experiment than to look back and see that simple errors could have been easily avoided. Errors could cost you time and delay your experiment or worse yet, mean you have to cancel the experiment before it is finished.

Experimental design types are typically selected based on the number of confounding environmental factors that may become involved. Since we want to learn if our experimental units differed because of treatment effects (and not the environment), we want to do everything in our power to control the environment and reduce any and all sources of variation that we possibly can. Will we control all of them? Surely not, but we can control most of them with planning, thought, and care. If you want to determine which experimental design to use, first consider the sources of variation that may be found in the environment of your experiment. If you have minimal variation and are comparing levels, rates, or product performance against each other, consider using a **completely randomized design (CRD)**, that we previously showed you in Chapter 3. If you have one source of external variation that needs to be controlled but still want to compare levels, rates, or product performance, consider a **randomized complete block design (RCBD)**. If there are two sources of external variation, consider using a **Latin square design (LS)**. There are other advanced designs that allow for mixing and matching of additional treatments based on adding and pairing other treatments (split plot, split block) but due to their nature are not included in this text. See the Appendix for a description of the 2-way factorial arrangement. This arrangement is very typical of agriscience and biological studies.

Additional designs can be used for qualitative research and we see these in surveys or questionnaires. These can be designed in a **pretest/post-test** fashion and take into account the differences between initial attitudes, thoughts, and knowledge that a sample population has before applying a treatment. The attitudes, thoughts, and knowledge from that same sample population are compared to the pretest data after the treatment has been applied. The differences between the "before and after" are analyzed.

Each of the first three designs will be discussed and explained here using several agricultural examples. Then in Chapter 6, you will be shown how the data will be analyzed using data sheets that follow the formats of the experimental designs found here. The point of this book is to show you how to develop a research project, create experimental objectives and an experimental design, organize your materials and experimental units, create your data sheets to collect data, randomize and care for your treatments during the experiment, and how to collect the data and analyze the results. If you correctly match the experimental design to the data sheet to the experimental arrangement and analysis, this will improve your chances of determining if there are statistical differences between your treatments which will lead to obtaining credible results. Now let us get back to the experimental designs.

Completely Randomized Design (CRD)

If we have a homogeneous group of experimental units that we wanted to include in an agriscience experiment (e.g., F_1 pepper seedlings or piglet siblings) that are being raised in an enclosed or very protected environment (e.g., greenhouse or farrowing house nursery), many sources of variation are already greatly reduced. Controlled variables in this design would include temperature, humidity, irrigation or water availability, reduced exposure to pathogens, and consistent feed and nutrient supply. With such a controlled environment already a part of the experiment, chances are pretty good that our experiment taking place in the farrowing house or greenhouse would expose any treatment effect differences and we would be pretty confident that those differences were due to those treatments applied to the pepper plants or piglets and probably nothing else.

Again, can we be absolutely sure the differences were only due to treatment effects? No, we are never 100% sure that our results came from treatment effects, but we need to have some confidence that we did the best we could to control experimental error. If the experimental error (random error) is really high, it might lead us to believe we found significant differences between our treatments when in reality we did not. Random error (also called human or experimental error) is caused by inaccurate measurements or carelessness and can reduce or increase the chance of showing significant results in the data, depending on how far off our errors are. The error could go either way, influencing the results from the data to either be significant or not.

Systematic error is another type of error comes from having a piece of equipment work improperly or it could be something that affects the treatments throughout the duration of the experiment, causing biased results. Both random error and systematic error can be reduced with careful planning, attention to detail and most importantly, by you looking at your experimental units often. Do you notice any problems? Does anything seem out of the ordinary? If you work with several lab partners, take some time to agree on measuring techniques and procedures so you have precise and accurate results. As you are looking at the growth parameters you have chosen, be aware of both random and systematic error with regard to your measurements. In statistics, claiming we have found a significant difference between treatment results when we really did not is called a **Type 1 error**.

Type 1 errors occur when one rejects the null hypothesis (accepting the alternative hypothesis that at least one of the treatments is different than the rest, statistically speaking) when one really should not have. An example of what contributes to a Type 1 error would be an experiment where one wanted to apply different brands of fertilizers on pepper seedlings to test which fertilizer brand promoted greater plant growth. If we allowed some source of variation to enter our experiment, it might affect our measurements. If we calculated our means incorrectly or computed some other statistical output incorrectly (or somehow included bias toward or against a particular fertilizer brand), we might find in our data differences between the fertilizer treatments when, in reality, there were no differences. This Type 1 error would lead us to report that one brand was superior to the others when, in reality, it was not. Math errors, bias, and other sources of error (e.g., using experimental units that were not statistically the same in the beginning) may lead to making a Type 1 error, stating there were differences in treatments where there really were not.

CRDs can help set up simple experiments because they allow us to be able to compare the results occurring from various types of a product (e.g., fertilizer brand, calf milk replacer, or variety of carrot) or can compare levels or rates of a product (e.g., rates of fertilizer or seeding density in pastures). Either way, the experiment is using one independent variable (i.e., the fertilizer) with several controlled variables (e.g., one variety of plants, same pot size, and same soilless media), and potentially several dependent variables (dry shoot weight, height, fruit yield).

As seen in the Figure 3-4 from Chapter 3 where plants were placed on the bench, all plants had an equal opportunity to receive the various fertilizer treatments but it was the luck of the draw that determined which plants were assigned to each treatment and where they ended up on the greenhouse bench. Since the experimental plants were all the same variety (or breed or age in livestock, like the piglet siblings), that reduces the chance of variability between them, which reduces the chance for experimental error. Experimental error takes away from the "true value" we are trying to determine, that is, what the true measurement is of our experimental units, in this case, the flowering plants grown under varying levels of fertilizer.

A CRD only works if you have a homogeneous group of experimental units in a controlled environment with minimal confounding variables. Controlled environment means that all experimental units have equal access to or receive the same controlled variables. It does not mean that the experiment needs to be in a greenhouse, growth chamber, or other enclosed structure, but that would surely be to your advantage.

In plant studies, seedlings of one variety can be started easily from seeds or cloned plants can be propagated from a single stock plant. In biology, mice are usually used as experimental units since they easily reproduce large numbers of offspring. Many biological and genetics studies use *Arabidopsis*, a fast-growing plant that can go through its life cycle in a matter of six weeks, and geneticists also find yeast a handy organism for genetic studies since it grows very quickly and can be genetically modified easily. Few people have any complaints about modifying yeast in research.

Animal scientists have a tougher time when developing their experimental designs since they try to use livestock of similar gender, age, and background, depending on the research. This may be harder to coordinate, may be more expensive, and can be a limiting factor when gathering replicates. Remember, we want to reduce the chance of experimental error and increase the chance that our treatments' results were not due to random chance but to treatment effects. If the experimental plants, yeast, or livestock are homogeneous in nature and type, chances are better that differences in the dependent variables (e.g., weight, ADG, height, yield) occurred due to the treatments they received and not because they were a larger animal or faster-growing plant from the beginning.

Another factor to consider when developing a CRD is the area in which the experimental units will be housed. The greenhouse bench, pasture, field, orchard, or nursery must all have the same controlled variables, that is, they all should receive the same amount of light, water, have the same temperature, same soil type, and same feed access. If any of these are not consistent across the test site, a CRD should not be used. But do not worry; there are other options for experimental designs that we can use if the testing site is not homogeneous, and we will get to those shortly.

Toys or small objects can help you learn how to assign treatments to experimental units in a CRD or other designs and what those units might physically look like at your experimental site. Lego® bricks or other construction blocks are handy toys to use in experimental design models because they have both plants and animals that can be color-coded with colored bricks (treatments), are durable, and can be modified when you are brainstorming with your research group.

To design a simple CRD with Lego® bricks having two treatments (time-release fertilizer versus liquid fertilizer) in greenhouse foliage crops, you would take a group of Lego® trees, divide the trees into two groups, and "assign" half to a time-fertilizer treatment (Osmocote®, for example) by attaching the Lego® tree to one color of brick, and "assign" the other half of the Lego® trees to the other treatment, a liquid fertilizer treatment (using Miracle-Gro®, for another example) with another color of Lego® brick. Half of the trees would be attached to red bricks and the other half would be attached to blue bricks.

The null hypothesis would be there will be no differences between treatments; each treatment would result in the same height (the dependent variable). The alternative hypothesis would be there would be a difference between treatments (one treatment would outgrow the other, resulting in taller plants). I recommend that you go through the randomization activity each time with your experimental criteria and it should help you to set up a map of your experimental units as if they were to be included in an experiment. This "dry run" helps you verify that you have enough space, materials, and plants for what you have planned in your objectives. It will get you ready for the real thing.

If you were growing 30 young weeping fig trees (*Ficus benjamina*) for this example and wanted to randomize them on the greenhouse bench with both treatments, you would need to separate the trees into two groups to receive each type of fertilizer (like in the Lego® tree example mentioned previously). You need to decide ahead of time where they will go into the treatments and where on the greenhouse bench (perhaps the first five selected from a random drawing are assigned to the Osmocote® treatment and the next five are assigned to the Miracle-Gro® group, alternating until all 30 are assigned to each of the two fertilizers). Using the information on randomization from Figure 3-4, you would want your design to look something similar but with only 2 levels of 1 independent variable (2 types of fertilizer) with 15 replicates/treatment.

Randomized on the greenhouse bench, your experiment might look something like Figure 5-1. If you take two different highlighter pens and color the one's on Figure 5-1 with one color and the two's with a distinctly different color, you can see they could be randomized and somewhat mixed up. Is it possible to have all of one treatment in a row? Sure, but if you randomized your plants into the two treatments to eliminate bias, you did your best to reduce experimental error. If they end up all in one row in the front, back or side, that is just the way it is. It was just the luck of the draw that they ended up where they did and no one can fault you on that.

The point of this experiment is to see which fertilizer increases plant height. We need to have 30 weeping fig trees available that are the same age, same species (and variety), and preferably the same height to begin with. We would take initial heights to make sure there were no significant differences between the trees prior to the beginning of the experiment. Do you remember what **significance** means? Significant results do not mean that they are important. What it means in statistics is that our data were shown to occur other than by chance, meaning, the results most likely occurred due to treatment effects.

Is that not what we are after in experimental research? Do we not want to see differences between treatments? Sometimes, and that depends on what you are studying. What if you have two products that supposedly work the same (fertilizers, for example) but costs for each vary greatly. It would be great to know you could use both products interchangeably and get the same nutritional benefit for your plants from either product. Of course you would probably select the fertilizer that costs less

(Top, north edge of bench)

2-4	1-4	1-15	2-2	2-3
1-2	1-14	2-15	1-5	2-5
2-6	1-1	1-13	2-7	1-6
1-10	2-8	1-3	2-14	1-8
1-9	2-12	2-13	1-7	2-10
2-11	1-11	2-9	1-12	2-1

(Bottom, south edge of bench)

Figure 5-1. Placement of each replicate on a bench for 30 plants (15 replicates/fertilizer treatment) in a CRD of two treatments.

to save money in the long run, and it is hoped that that is obvious. We can determine statistical differences with computer statistical software and spreadsheet programs, which we will show you later on in Chapter 6.

Remember that in our weeping fig tree example we want significant differences between our treatments at the conclusion of the experiment but not at the beginning of the experiment. Since we need homogeneous experimental units to conduct a CRD, we need to test them to see if they are in fact "the same" significantly from the start. Is this not along the same lines of what we were discussing about significant differences a moment ago? Yes, so to get this information, we need to take the initial heights of all plants used in the experiment (in centimeters with rulers that are of the same brand/type to increase precision) and enter these data into a statistical software package to have it determine if our initial data are significantly different. If the data are not significantly different, then that is like saying our initial data are mathematically/statistically the same, they are interchangeable, they are equal. That is exactly what we need, a group of initial data that is homogeneous in nature and is found equal or "the same".

So, a CRD allows us to assign different levels or rates of treatments or types of similar products to experimental units (units that are homogeneous in nature) that all have access to the same controlled variables. What would another CRD look like when comparing rates of a product? Here is an example that shows how to compare different levels of a single type of fertilizer to determine the best rate (both in terms of amount and cost).

Suppose you are interested in growing tomatoes in the summer so you read the horticulture literature and discover that various amounts of fertilizer have been shown to be effective in growing different species of vegetables. Since these fertilizer levels (rates of N in ppm) apply to different crops and there is no consensus on a general recommendation, you are not sure which rate you should use to get your tomatoes up and growing. You do not want to use more fertilizer than you have to since 1) you do not want to increase fertilizer run off into irrigation water and 2) you do not want to pay for more fertilizer than you need.

The first thing to consider when developing your tomato experiment into a CRD is to test the rates suggested for vegetables, in this example suppose there are seven different rates suggested in the literature. You know from previous discussion that several replicates are better than a few to find the "true meaning" so you choose to include five replicates/fertilizer treatment. Again, using more replicates helps reduce experimental error which it is hoped will increase your statistical power later on when you conduct your data analysis.

As you plan the experiment, review and consider developing an experimental layout using coding variables from Table 3-1 (in Chapter 3) and randomization methods to obtain something similar to Figure 5-1. Determine the total number of plants that you will need in this experiment. Take the replicates needed/treatment (five replicates) and multiply that by the number of treatments you will be testing (in this case, seven). A simple math calculation of 5 x 7 = 35 yields the number of tomatoes that will be included for this study. Figuring this out will help you determine the size of the greenhouse bench or other housing necessary for your experiment, and help you determine how many plants to grow, how many to prepare for, etc. Proper planning is vital to the success of any experiment. You cannot expect to "wing it" and get reliable results. A good rule of thumb in greenhouse foliage, bedding plant, or vegetable transplant research is to over plant by 20%.

Having extra plants on hand allows you to pick the most healthy and homogeneous specimens for your trial. If you know you need 35 plants, 20% of 35 is 7. Adding 7 to 35 is 42. You will need to sow at least 42 tomato seeds or buy 42 tomato plants to have on hand for this experiment. Again, it is always best to have more plants or other experimental units on hard to select from so you do not have to use smaller, less-than-desirable experimental units that lack consistency or homogeneity in the first place. If your experimental units are not statistically the same in the beginning, you cannot say for sure that your treatments caused them to be statistically different at final harvest. You will never really know for certain if any differences were due to initial plant size or treatment effects, and that will be a problem.

Back to the tomato fertilizer trial example, you would repeat the steps of randomizing the plants and placing them on a greenhouse bench that were covered in Chapter 3. Apply each fertilizer to each replicate in each treatment by hand (starting with the control and lowest treatments first going through the treatments up to the most concentrated) with a measuring cup to ensure each plant gets the same amount of liquid volume. Why should we start fertilizing with the lowest concentrations of fertilizer? If you apply them in this order (weakest to strongest) and you are in a hurry, you do not have to rinse the measuring cup in between treatments since the next fertilizer treatment is stronger. Got it?

Making Liquid Fertilizer Treatment Solutions

How would you go about setting up the treatments if you really wanted to try out this rate experiment using water-soluble fertilizer? To make fertilizer rates for this type of experiment (with nine treatments and five replicates/treatment), you would need the following supplies:

(10 each) 5 gallon buckets with lids

(2) Large mixing spoons

Miracle-Gro® fertilizer (or other water-soluble brand)

Measuring cups and spoons, assorted sizes

Scale

Access to water

Index cards, permanent marker, masking tape

Potting soil

Plant/pot labels

(1) 72 tray plugs or liners

4" (10 cm) plastic azalea or geranium pots (minimum 72)

The first thing you need to do is have the plants available in plug trays (have them ordered well ahead of time) or start seed to grow your own plugs (about 3–4 weeks ahead of time, depending on variety grown). How many plants do you need? You should have 20% more than what will be included in the experiment. Nine treatments x 5 plants/treatment = 45 plants. Add 20% (0.20 x 45 = 9), so you will need to eventually pot up 54 plants into 4" pots.

Once the plants are on site, confirm your treatments and make labels for your buckets with the index cards and markers. Make sure you label the lids in addition to the buckets, and line them up against the wall in order of concentration; it will make finding each treatment quicker and easier. The literature suggests 100–200 ppm N but many professional growers give their crops 200 ppm N to be safe. How much N (in ppm) is provided in home fertilizer products? This varies, some are as high as 900 ppm N and some are a third of that.

To begin making your fertilizer stock solution, from which you will make the rest of your treatments, you will need to read the package of water-soluble fertilizer. The guaranteed analysis of a fertilizer is the set of three numbers typically found on the front of the package, box, or bottle and is the percent of nitrogen, phosphorus, and potassium in that container. This information is required to be listed on the label by federal law. These numbers are the largest of the macronutrients that are supplied to plants through mineral elements either as gases, solids, or liquids. The three numbers specifically represent the percentage of N, P_2O_5, and K_2O in that package. In a product that has 24-8-16, for example, you would be getting 24% N, 8% P_2O_5, and 16% K_2O. Are you really getting 8% P and 16% K? Not so, P and K are listed as oxides in a **complete fertilizer** (a complete fertilizer provides all three numbers) while N is listed on an elemental basis.

If you take the atomic weights of P (31), K (39.1), and O (oxygen, 16), you can add up the overall molecular weights of P_2O_5 and K_2O, divide by the atomic weights of each element, and get the percent of P and K on an elemental basis. For example, how much P is in P_2O_5?

P_2O_5 has 2 P and 5 O, therefore P (2 x 31) = 62 and O (5 x 16) = 80, 62 + 80 = 142. Since the atomic weight of P in P_2O_5 is 62/142, this computes to 0.4366 P which is rounded up to 0.44.

You will typically read in the literature that there is 44% P in P_2O_5. What is the percent of K in K_2O? Can you work it out? It is 83%; confirm that you know how to compute the percent of K for future fertilizer or chemistry research.

For the fertilizer problem, you have many choices of fertilizers with various guaranteed analyses available at local nurseries and retail garden center chain stores. Select a water-soluble fertilizer product and make a stock solution; from it you will make your individual treatments. A stock solution is a very concentrated fertilizer solution that will be diluted down into the appropriate treatments. Using fertilizers with different guaranteed analyses will change the concentration of the stock solution each time, so knowing how to compute the stock solution will allow you to make the same treatments again even if you have to use a different fertilizer to make the stock solution. There are a couple of different ways to calculate the amount of fertilizer needed to make a fertilizer stock solution, and each method will use a different fertilizer guaranteed analysis to provide you with additional examples.

Example #1: Using 20-10-20 water-soluble fertilizer available in bulk, weigh and add 1000 g of fertilizer into a bucket or large Nalgene® bottle that has a tight lid or screw cap. Add 10 L of water and mix thoroughly with a large spoon. If 10 L of stock solution is too much to store (you know you will not need all of it), consider making only 5 L in the future.

We want to measure our treatments in ppm but have our measurements of fertilizer and water in grams and liters. Luckily, the conversion of ppm is mg/L (or $mg \cdot L^{-1}$) so we can convert without difficulty. Calculating a stock solution from grams and liters to ppm N is made easier if you follow the equation below. Remember that 20-10-20 contains 20% N, so 1000 g of fertilizer really contains 200 g N (1000 x 0.20 = 200). Watch your units:

$$\frac{1000 \text{ g fertilizer}}{10 \text{ L water}} = \frac{200 \text{ g N}}{10 \text{ L water}} = \frac{200{,}000 \text{ mg N}}{10 \text{ L water}} = \frac{20{,}000 \text{ mg N}}{1 \text{ L water}}$$

$$= 20{,}000 \text{ ppm N in this fertilizer stock solution}$$

So, mixing 1000 g of 20-10-20 water-soluble fertilizer into 10 L of water will generate a stock solution of 20,000 ppm N. This is far too concentrated to use in our experiment so it will need to be diluted down to get it into the appropriate treatments that we want to include in our study. Since we want to include the following treatments from Table 5-1, taking a tenth of the stock solution and adding it to enough water to make 1 L will make 2000 ppm N. If 2000 ppm N is made from taking 1000 mL of concentrated stock solution and adding it to 9000 mL of water (this is 10%), then a reduction calculation shows that 200 ppm N comes from adding 100 mL stock solution to 9900 mL water. From that, we can solve for the other concentrations with ratios or use of a calculator (Table 5-1):

Example #2: Using 24-8-16 water-soluble fertilizer in a small 1.5 pound box (available at most nurseries or garden centers), pour the entire contents of fertilizer from the box into a bucket or large Nalgene® bottle. This 1.5 lb box of fertilizer has a mass of 680 grams. Add 10 L of water and mix thoroughly. As seen in the previous example, remember that 24-8-16 contains 24% N, so 680 g of fertilizer really contains 163.3 g N (680 x 0.24 = 163.3):

$$\frac{\text{680 g fertilizer}}{\text{10 L water}} = \frac{\text{163.3 g N}}{\text{10 L water}} = \frac{\text{163,300 mg N}}{\text{10 L water}} = \frac{\text{16,330 mg N}}{\text{1 L water}}$$

$$= \text{16,330 ppm N in this fertilizer stock solution}$$

So, mixing 680 g 24-8-16 fertilizer into 10 L of water will generate a stock solution of 16,330 ppm N. Again, this is far too concentrated to use in our experiment so it will need to be diluted down to get it into the appropriate treatments that we want to include in this research.

Compare the treatment ingredients in Table 5-2 to those we had in Table 5-1. Since we want to include the following treatments as before (in ppm N), taking a tenth of the stock solution and adding it to enough water to make 1 L will make 1633 ppm N. If 1633 ppm N is made from taking 1000 mL of concentrated stock solution and adding it to 9000 mL of water (this is a 10% dilution), then a reduction calculation shows that 163.3 ppm N comes from adding 100 mL stock solution to 9900 mL water. From that, we can solve for the other concentrations as we did previously. Table 5-2 gives the ingredients needed using stock solution made from a 1.5 lb. box of 24-8-16 water-soluble fertilizer:

Using fertilizers that have different guaranteed analyses changes the concentration of the stock solution so remember to read the label of the fertilizer product that you are going to use and plan to make the appropriate calculations accordingly.

Table 5-1. Fertilizer rate solutions based on using 20-10-20 for the stock solution. Measure the amount of stock solution and add it to water in a bucket, creating that treatment rate.

Treatment (ppm N)	Stock solution (mL)	Water
0	0	10 L
50	25	9975 mL
100	50	9950 mL
150	75	9925 mL
200	100	9900 mL
400	200	9800 mL
800	400	9600 mL
1600	600	9400 mL
3200	1600	8400 mL

Table 5-2. Fertilizer rate solutions based on using 24-8-16 for the stock solution. Measure the amount of stock solution and add it to water, creating that treatment rate.

Treatment (ppm N)	Stock solution (mL)	Water
0	0	10 L
50	30	9970 mL
100	61	9939 mL
150	92	9908 mL
200	123	9877 mL
400	245	9755 mL
800	490	9510 mL
1600	980	9020 mL
3200	1960	8040 mL

What does the literature say about optimum N fertility? It varies by crop, method of irrigation, and whether the crop is grown in field soil or potting media. This is an interesting study and may help reduce the amount of fertilizer we use in commercial agriculture and by home gardeners. Did you know home gardeners are known for over-applying fertilizers and other home chemicals because they think "if a little is good, more is better"? Think about that.

Randomized Complete Block Design (RCBD)

RCBDs are used extensively in agriscience research due to the nature of variability found in soils in fields, pastures, orchards, and livestock diversity. The old saying about designing experiments is "randomize what you can, block what you cannot." When there is variation in a pasture, field, garden plot, greenhouse structure, or nursery, it makes sense to block the areas and ensure all treatments are represented within each area of variation. This is called **blocking**, creating homogeneous experimental units.

What if we have a situation where a homogeneous group of experimental units (e.g., F_1 pepper seedlings or piglet siblings as seen previously) are being raised in an environment that may have some source of variation? Perhaps what we thought was a controlled variable in our experiment changes by location, such as temperature or soil type. If this is the case, this controlled variable is no longer controlled and needs to be accounted for.

Another situation may arise when we do not have homogeneous groups of experimental units (most likely seen in livestock). Some animals weigh more, some less. We may have a question about breed of livestock, does breed play a role in weight gain or milk yield? Gender in livestock may be a problem; do males gain faster than females? Blocking animals into weight categories and treating each weight block will help eliminate bias in livestock research.

Blocking creates a situation where each treatment receives or is grown under each area of variation, therefore maximizing differences between blocks, and reducing the chance of experimental error. If there is a greenhouse with benches in different locations, it is possible that one bench could be warmer than another; it could get more sun, or could get less irrigation water. We would place replicates of our peppers from all treatments, for example, in each block (location). Blocking by location would ensure that all pepper treatments are placed in each area that may have the variation, resulting in that variation being applied to all treatments.

Again, minimizing the variation would increase the chance that our experiment would expose any treatment effect differences and we can be confident that those differences were most likely due to the treatments applied to the pepper plants and (it is hoped) nothing else, especially the differences in the greenhouse, soil, etc.

Figure 5-2 shows what we might consider blocking if we wanted to test different fertilizer treatments (A–D) in peppers but had a long greenhouse bench that received different amounts of sunlight each day. Each treatment would be a fertilizer rate and each block would be a replicate. In this example, there are only 4 plants/block, totaling 16 plants in the study (Figure 5-2). Each block would take each area on the bench that receives different amounts of sunlight into account. Perhaps Block 1 in Figure 5-2 gets most of the sun; we would be able to determine if these treatments in Block 1 are growing more as a group. We are trying to see if there are any trends in the data by block. Are there any blocking effects? Statistical software can determine this for us, which we will see in the following chapter. In a livestock example, perhaps we are testing the palatability of new dairy feeds and have different calf weights. Figure 5-2 would have each group of animals blocked by weight. Block 1 would be the lightest calves, Block 2 would be calves of light-medium weights, Block 3 would be medium-heavy weight calves, and Block 4 would be the heaviest calves.

RCBD is a good design to use if you need to determine if there are blocking effects but with such few plant or animal numbers, do we really have a good opportunity to see if there are treatment differences? A modification to the RCBD is to subsample, that is, include additional experimental units in the design.

Subsampling is demonstrated in Figure 5-3 by adding numbers 1, 2, and 3 which represent three additional replications of each treatment within that block. It might be confusing to think that a block is a replicate and then to see additional replicates within a RCBD subsampling treatment. Keep a RCBD simple by calling homogeneous groups of experimental units "blocks" and additional units within each treatment "replicates" (Figure 5-3). In Figure 5-3, Block 1, we see three plants or three dairy calves in each treatment in each block (A1, A2, and A3). This provides extra experimental units that will help reduce error and increase the chance of detecting treatment effects. Having three calves/treatment makes more sense since you get more data to compare in your analysis.

Could these blocks be repeated at other sites (at other feedlots, for example)? What about using additional fields or other greenhouse buildings? They could be and typically they are; the more replicates you have, the more data you can analyze, and the greater chance you have of determining statistical significance if it exists. The work and time invested in developing experiments usually includes testing at additional sites to include more replicates of each treatment. The important thing to remember is

(Top, north edge of bench)

Block 1	Block 2	Block 3	Block 4
A	D	B	D
B	B	C	C
C	A	D	A
D	C	A	B

(Bottom, south edge of bench)

Figure 5-2. Placement of each block (replicate) on a bench for 16 plants (4 replicates/fertilizer treatment) in a RCBD of 4 treatments (A–D).

(Top, north edge of bench)

Block 1	Block 2	Block 3	Block 4
A1, A2, A3	D1, D2, D3	B1, B2, B3	D1, D2, D3
B1, B2, B3	B1, B2, B3	C1, C2, C3	C1, C2, C3
C1, C2, C3	A1, A2, A3	D1, D2, D3	A1, A2, A3
D1, D2, D3	C1, C2, C3	A1, A2, A3	B1, B2, B3

(Bottom, south edge of bench)

Figure 5-3. Placement of plants within a RCBD with subsampling. Each treatment has three replicates/treatment in each block as might be found on a greenhouse bench.

that RCBD blocks experimental units into homogeneous groups (all blocks have the same treatments being applied) and you can replicate this as much as your budget will allow.

Latin Square Design (LS)

A Latin square experimental design is also used in agriscience research and expands the idea of RCBD by making an allowance for two sources of variation. The RCBD removes one source of variation, the LS removes two. This is helpful in agriscience experimentation since it is very difficult to have plants and animals that are homogeneous in nature and are grown/raised in optimum environmental conditions. In our previous dairy calf RCBD testing dairy feed, we had calves blocked by weight. In a LS design, we could block by both calf weight and time on new feed, testing the palatability of feed brand. We also see this in food science studies when new products are undergoing taste-testing. Suppose you wanted to test some new trail mixes to get some ideas of consumer preferences and purchasing habits. This is no different than testing palatability of dairy feed. We need to have items to test (feed or trail mix) and consumers to taste them. Each food taster (or calf) has the opportunity to try each product in the test.

A problem with food or feed tasting is that there needs to be a time delay (or palate cleanse) in between each rotation of taste tests to eliminate "leftover" or carry-over effects. This is common in taste tests where a tester still has thoughts or feelings about the previous item tasted. This can go for the product or against it, depending on the previous experience had by the tester.

Another problem in taste testing can come from the actual time of day an item was consumed or the order in which it was consumed. Was the taste tester hungrier at the beginning of the trial, therefore the first samples tasted better? Was the tester getting full or tired by the end of the taste testing? Having a plan to eliminate such variability is helpful and Figure 5-4 shows an example of four judges/consumers/animals involved with a taste-test trial. Blocking by both tester (human or livestock) and time consumed in a LS design will (it is hoped) remove any bias. Figure 5-4 shows that each consumer or judge tries out each treatment and that all treatments are being tried during each round of testing.

Each round of testing could be a specific block of time assigned to each group of livestock. Perhaps we want to test the usefulness of four automatic watering spigots for hogs in Figure 5-4. The LS design removes the variation between the hogs themselves and also the variation of having access to the four different water spigots in different order. The first round would last for two weeks, with a three-day buffer where the hogs get used to using the new water spigot. Then the next type of spigot is installed (the previous one is removed), there is an adjustment period so the animals get used to the new water source, and the amount of water consumed (lost, used, whatever it is you want to know about water spigots and their use by hogs!) is measured. After the trial period is over, there is another three-day adjustment period after the new spigot is installed and after another two weeks of collecting data, the third spigot is installed, the pigs use it, measurements are recorded, and so on.

It is important in this study to remove any variation caused by the rotation because climate can play a role in how much water is consumed: it may get hotter later on in the trial, more water is lost due to greater consumption, and water may be lost due to playfulness (hogs like to play in water, really!).

A horticulture LS example would include one where we have both soil fertility and irrigation application variation in student vegetable gardens. Suppose there were four legume cover crops grown in the gardens the previous year and we have four sprinkler irrigation systems available to the students in the regular semester. We want to test four new broccoli varieties (Figure 5-5, items A–D) to see if they will grow well and produce in the student gardens for possible use during the next semester under our growing conditions.

Each garden plot grown in the cover crop row from Figure 5-5 would have its own sprinkler system (1–4) and would grow one of four types of broccoli. Again, if you are worried about replicates and want more data, subsample with three broccoli plants per row/column intersection.

Pretest/Post-test Designs

Sometimes there are opportunities to survey people about their preferences of one product over another or what people have learned over time. As seen in the LS design discussion, there is a lot to learn about tastes, food appeal, and consumer likes and dislikes and the LS design reduces the variability of having different consumers participate in the experiment. Taste tests can be subjective in nature so the more you can make the test objective, the greater chance you have of reducing experimental error.

Another experimental measurement that can be somewhat subjective in nature is collecting survey data through attitude and knowledge tests. Surveys are subjective test instruments because people's attitudes can change quickly from one day to the next, but we can write survey questions where we can expect certain answers in some situations. Many of us have taken surveys that follow a **Likert type scale**. These

	Consumer 1	Consumer 2	Consumer 3	Consumer 4
Round 1	A	C	B	D
Round 2	C	A	D	B
Round 3	D	B	A	C
Round 4	B	D	C	A

Figure 5-4. Placement of each of four trail mix blends (four replicates/trail mix treatment) in a LS design with a panel of four judges (consumers) each trying the trail mixes (A–D).

	Sprinkler 1	Sprinkler 2	Sprinkler 3	Sprinkler 4
Cover crop 1	A	C	B	D
Cover crop 2	C	A	D	B
Cover crop 3	D	B	A	C
Cover crop 4	B	D	C	A

Figure 5-5. Placement of each of four broccoli varieties A–D (and subsample with three replicates/broccoli variety) in a LS design with four sprinkler lines applying water to the broccoli plants (A–D) grown under residual legume cover crops.

items typically include five answers (items) including strongly agree, agree, neutral, disagree, and strongly disagree. Since we might have an idea how people feel about topics (or should feel, if they have a conscience), we can compose our questions so that any of those five answers possible. We see a lot of these types of surveys and questionnaires in food science development and educational studies. Faculty evaluations usually include questions about the course that are answered with Likert type scale items. The instructor of this class always reads teaching evaluations to improve the quality of the course and gather new ideas so make sure you submit your evaluation when it is time to complete them.

Survey instruments need to be tested for **reliability** and **validity**. Reliability tells us that the test instrument gives similar results each time it is given to a subject (this is a measure of precision) and validity checks that the instrument is testing for what it was intended to test. Both of these tests can be performed with statistical software programs and it is a good idea to always check these tests before doing any large-scale research survey.

Knowledge tests are also given as pretests and post-tests to assess an individual's initial knowledge base and then given again after a treatment has been applied (e.g., a lesson or unit of study was taught or an activity was experienced). Comparing knowledge (or attitude) pretest data to post-test data can tell us several things. First, we can look at the overall differences in test scores between the pretest and post-test. Was there an increase in the overall test score? What was the overall difference? Is there improvement being made in specific areas?

This is a useful tool that can measure if lessons or activities are beneficial to the test population or if we want to change an old standard that we have been using for a long time and need input. Will an activity be missed if we stop using it? Another consideration with pretests and post-tests includes looking at specific questions in the test. Can we pinpoint where we need to make improvements on a specific topic or objective?

We can assign coding variables to the Likert type scale items to quantify the qualitative data like we saw in Table 3-1 from Chapter 3. This will help us calculate if there is a difference between attitudes after a treatment is given.

Pretest and post-test knowledge tests can be compared by individual test question or by overall test score. We can also analyze improvements by age, gender, year in school, or other descriptive statistics that we gather from a biographical information section. Do not hesitate to ask about a participant's background if you think it could influence your dependent variables. The information you gain from biographical data can help explain some of the responses you get. This in turn may lead to further research studies.

Using Surveys to Develop Snack Foods

In food science we have taste tests to develop new products or blends of products. Classic examples of food tests include blends or mixtures of several components such as sausage patties, fruit juices, trail mixes, frozen vegetable blends, and fresh salad mixes. In fruit juice development we often see blends of orange juice with pineapple juice and banana juice. Having survey taste tests with "unknown" fruit juice blends (e.g., Juice A, Juice B, and Juice C) allows us to score the appeal (color, flavor, and aroma) of a product to the general consumer before it goes out to market. Using unknowns, we have reduced bias since the taste-testers do not know what brand or combination of juices it is they are testing; therefore this reduces the chance of adding personal bias to the study. We could use a Likert type survey to gather additional information about our products and how people feel about them or get their ideas about developing new products.

The California agricultural industry produces a huge assortment of fruits and nuts for the consumer. While great to eat plain, we have found that mixing fruits and nuts (and chocolate candies too) has given a big boost to fruit and nut sales through the promotion of trail mixes as a snack food. While we will continue to sell bulk fruits and nuts for other uses, having an alternative market for these products is beneficial to the grower, since it provides additional marketing and sales strategies to move product to untapped purchasing audiences.

Mixing agricultural commodities together to create a new product to reach additional customers is called **value-added marketing.** Products that are marketed through value-added merchandising may open doors to non-traditional customers and clientele that might not otherwise be reached. We see this all the time with fresh salad blends, frozen mixed vegetables, and mixed fruits and nuts, such as those found in trail mix. What are the main ingredients that you typically see in common trail mixes? If you go to a convenience store and look at different snack mix ingredients, you will probably find many that have peanuts, raisins, and chocolate candies as their main components.

There are many other ingredients that can be found in trail mix blends (e.g., almonds; walnuts; cashews; sunflower seeds; dried fruit such as cranberries, apples, and bananas; cereal; pretzels; miniature marshmallows; etc.), and finding the right combination that provides taste, texture, and eye appeal simultaneously is a constant question for food scientists and food marketers. Each ingredient must contribute specifically to the mix, whether it is taste, texture, aroma, or color. The right combination can make or break a sale at a convenience store, and the snack food industry depends heavily on taste-testing to determine what products go out on the market. Fruit and nut growers are particularly interested in food marketing so, depending on your major, this might be something worth pursuing for your career.

How would you go about designing a visual appraisal and taste test to develop new trail mix blends? Think about this in regards to the types of ingredients that you like to eat; the aromas, colors, and textures of those ingredients; and how much each ingredient costs. What are you willing to pay for trail mix and who would your target consumer be? What about packaging materials? What would be an attractive retail package? Agribusiness companies and marketing firms work on new packaging and advertisement plans all the time, constantly creating new materials that are attractive, eye-catching, and lend excitement to the product. Research in agriscience is not just about whether one crop will grow better than another or if one type of livestock feed will contribute to greater average daily gain, it is way beyond that. It involves the food marketing industry as well as agribusiness (finance and investment, personnel management, and accounting). What you do with your knowledge about agriscience research can open many doors for you and your career, no matter what field you are in (plant science, animal science, horticulture, forestry, meat science, etc.). Knowing how to look at data and make decisions from those data will separate you from your competition when you seek employment. Keep an open mind and let experimental design work for you to help you reach your career goals.

In summary, there are many ways to set up experimental designs to help eliminate bias and reduce variability. Experimental designs can help us make better choices and can help teach us if we can exchange products or techniques interchangeably. Through the use of a CRD, RCBD, LS, surveys, and pretest/post-test designs, attempts are made to include randomization of treatments to experimental units, assign treatments to homogeneous blocks, and utilize two blocking factors, all of which are designed to reduce experimental error and determine if significant differences occur between treatments.

In this next exercise, you will see practical applications of the experimental designs discussed in this chapter and will create models of some of the designs in lab. Think about what you read in commercial advertisements and in sales documents, did they use the correct types of designs to obtain their results? Be critical of their findings and be skeptical of products' claims that may seem to be "too good to be true."

6 *Data Analysis and Interpretation*

— Managing Raw Data
— Types of Data Analyses
— Using a p-value to Determine if Our Data are Significantly Different
— Experimental Arrangement Examples Showing Significance
— CRD Analysis with Excel and JMP® SE software
— Using the Tukey Multiple Comparisons Analysis to Determine if Data are Statistically the Same as Other Data Values
— Using Correlation Coefficients and Linear Regression to Correlate Variables and Make Future Predictions
— Using X^2 (chi square) to Determine if Count Data are Significant
— Exercise 6-1. Modeling Experimental Arrangements

Managing Raw Data

Now that your experiments are developed and you have thought about how you are going to collect data, it is time to think about how you are going to record your data points and deal with them once you have them. How are you going to analyze your data? What can all of the different types of analyses tell or suggest to you? Do they all mean the same thing?

Before you start to wonder if you have to get a statistics book, do not worry, we will help you here by showing you what you need to know and how some statistical software programs can do the work for you. You need to think about one thing after the statistical analysis: are the data significantly different? That is going to be important when interpreting your results.

Do not forget, you need to collect your data in the beginning of the experiment (initial data) and at least one or two times throughout the experiment. At the end, you will need to collect final data, then harvest or terminate your experiment. Will you always need to collect data throughout the experiment? No, but it is always better to have the information and look for trends along the way than to get to the end and wonder if you made a mistake in your measurements, if something was wrong from the start, or if there truly was some new phenomenon occurring.

If you are using livestock in your agriscience experiments, you can either return the experimental animals to their original herd or flock, or you can sell them at market. Plants are usually terminally harvested to get dry shoot weights, so remember to have your camera and extra batteries nearby to take final pictures well beforehand.

Some of the final data that you can collect will be determined by the type of plant species you are using (Table 6-1) or animal species (Table 6-2), their ages, and their uses. There are several options of dependent variables to measure for both plant and animal (or bacterial, fungal, or physical) experiments, depending on what you are researching. Make sure you remember to record your data to the nearest tenth of the unit of measurement (0.1) each time to ensure accuracy and precision. Some of the units are subjective (e.g., flower number) and change throughout the study so take care when selecting your units of measurement.

Types of Data Analyses

The point of data collection is to see if there are significant differences between treatments. If there are (and it is determined through the use of some statistical software or by hand with a calculator) you will probably be given a test statistic. If the test statistic is significant (usually predetermined less than or equal to an $\alpha = 0.05$ level) then the null hypothesis is rejected. That is pretty much what statistics tell us. They tell us whether or not to reject the null hypothesis and that is why you need to have some knowledge of statistics when you develop your experiment.

Table 6-1. Types of experimental plant data and units of measurement.

Dry shoot weight (g)	Presence or absence of flowers, bulbs, corms (count data)
Fresh shoot weight (g)	Time to flowering or fruiting
Height (cm)	Time to germination
Leaf number	Tissue test (% mineral nutrient content)
Leaf area (cm^2)	Soil test (% mineral nutrient content)
Flower number	Tree girth (cm)
Fruit yield (hundredweight, tons)	Presence or absence of pests (insects, fungi, mites, weeds, bacteria, large herbivores)
Percent germination	

Table 6-2. Types of experimental data and units of measurement used in livestock research.

Average daily gain (ADG, pounds)
Milk yield (pounds)
Percent butterfat
Wool yield
Birth weight (pounds)
Litter size
Percent lean tissue
Percent backfat
Percent intramuscular fat
Time to pregnancy
Percent synchronized
Dressing percent
Presence or absence of pests (insects, fungi, mites, worms, parasites, bacteria)

First, let us realize that you probably do not have the ability, resources, or time to measure every single experimental unit found in the **population** (all living experimental units in that group) you are studying so the units you selected to participate make up a **sample**. Sample sizes need to be large enough to allow us to detect differences as a true representation of that population and they need to come from a normal population. What is a **normal population**? It is a group where there are a few outliers toward each end of the spectrum and a large group in the middle, the typical bell-shaped curve (Figure 6-1). The mean would be directly down the middle, which in Figure 6-1 is approximately at 17 on the x-axis.

Normal data is not skewed one way or another; it is somewhat balanced with a large rise in the middle and tapered on each end. The sample size needs to be large enough so any differences between the individuals in the treatments can be detected, and that is why we need to include as many replicates in our treatments as possible. It has been suggested that a good sample size is at least 30 experimental units.

Since we know that we want our data to be significant, that is, we want to be able to reject the null hypothesis (which usually states there is no difference between treatments), we need to know where to find that information. This is where the use of statistical software comes in. **Depending on what we want to know, there are several statistical tests and methods we can perform to tell us if our data set is 1) significantly different (probability, p-value), 2) statistically the same as other data values (multiple comparisons, Tukey), or 3) can be correlated or associated with another variable.** Let us go through each scenario and example of each method to see if we can set up data analyses and data sheets to match each question we want answered about our data. We will learn about p-values, Tukey multiple comparisons, and correlation between variables.

Using a p-value to Determine if Our Data are Significantly Different

Is there a difference between the data in our treatments? This is probably the most common question all researchers want answered. While it sounds simple, there are several ways to go about answering this question. Ask yourself, what are we comparing; it is data from only two populations or from two or more? If it is only two groups, we can use a **t-test**. A t-test calculates whether the means obtained from two groups are statistically different. If there are two or more groups, we would want to use an

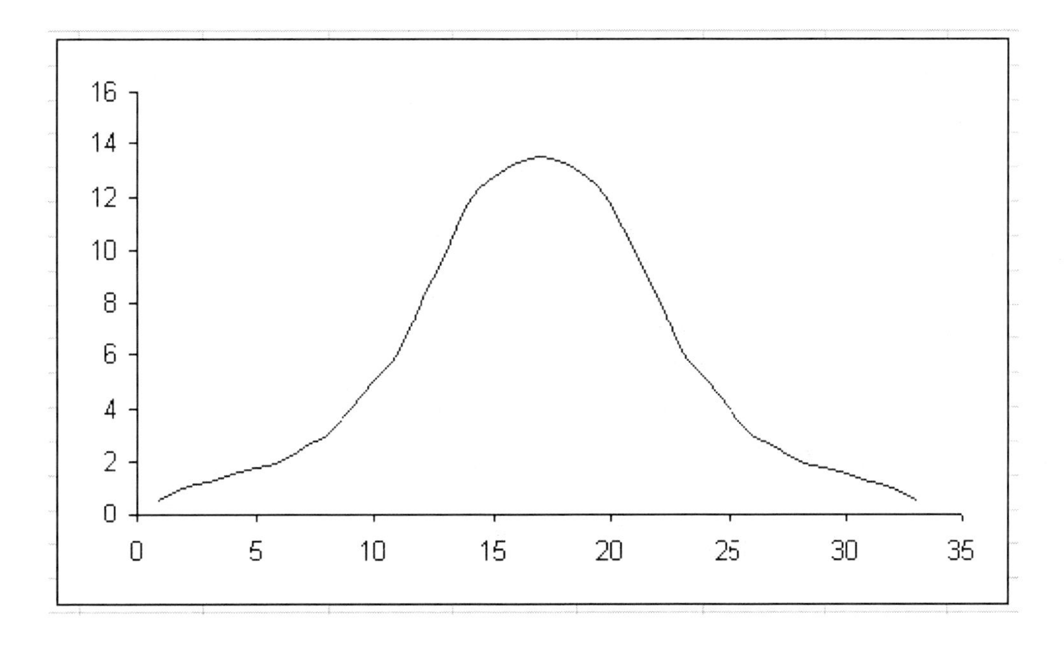

Figure 6-1. A typical bell-shaped curve from a normal population.

analysis of variance (ANOVA) test which tests for differences between the variances of two or more groups. An ANOVA will do what a t-test does but not vice versa. ANOVA can handle the data from several treatment groups but a t-test will only compute the difference between two means. ANOVA will be a key procedure in many of our experimental designs.

Are we comparing count data? Count data are different than means or averages, they are actual counts, numbers that represent if something was done or not, numbers of one category versus those in another. In seed germination trials, seed either germinates or it does not. In dairy cattle, cows are either pregnant or open. They are discrete choices and it must be one choice or the other. The statistical test for count data is typically the **chi square test (X^2)**. Chi square computes the probability (a **p-value** or probability value) whether the counts are statistically different, that is, whether there is a difference between the treatment groups.

The test statistic in an ANOVA for two or more treatments is the F ratio. The F ratio also computes the probability (p-value) of data occurring by random chance if the null hypothesis is correct. Do we want our final results to occur by random chance if the null hypothesis is correct? Certainly not, we want our results to occur due to treatment effects! It does sound a bit counterintuitive though, so try to focus on the part where we want to have our data resulting from treatment effects.

Before we set up our experiment we need to decide what risk we will take of making a Type 1 error (where we would reject the null hypothesis when in fact we should not). This risk is called the alpha (α) and is predetermined prior to running the data. Most experiments have an $\alpha = 0.05$ but many researchers in the social sciences use an $\alpha = 0.10$. Working with human responses can be quite challenging so in order to account for the subjective nature of working with people, higher alpha values are allowed.

How the p-value works is that it must be equal to or less than the pre-set α in order for significance to be observed. For example, a computed p-value of 0.04 would be significant at an $\alpha = 0.05$. It would be significant at $\alpha = 0.10$ but not at an $\alpha = 0.01$. Our statistics tests (t-test, X^2, and ANOVA) all come down to the computed p-values and this p-value is compared to a predetermined alpha to see if there is significance in our data or not. This p-value of 0.04 would be telling us that our data:

1. Could occur randomly by chance 4 out of 100 times or
2. We can interpret the p-value as 96 times out of 100 the differences in our data occurred due to the treatment effects (that the treatments were the reason behind our results).

Always look for the p-value but consider this, do you always want to find significant differences between data sets? Remember the experiment mentioned in Chapter 4 between the Miracle-Gro® and Osmocote® fertilizers? We wanted to know if we could use either one interchangeably, could one replace the other without there being any differences other than cost?

For this experiment, we would want our data to be statistically the same, that is, that there is no difference between the fertilizer treatments. We would not want to reject our null hypothesis that there is no difference between these two fertilizer brands but in fact would **"fail to reject"** the null hypothesis with a p-value of 0.051 or greater. Wow, what a mouthful! Yes, that is the way we say it, we fail to reject the null hypothesis if the p-value is greater than your pre-determined alpha. In this case that is a good thing, if we want to use the fertilizers interchangeably and the p-value that came from the statistical analysis supports our question of whether we can do this, switch fertilizers.

Please note, the results we gain from statistics tests can only support or refute your hypotheses, they never prove anything. There is no such thing as proving anything in science since we will never know the data from every single experimental unit in a population. What if we had every member of a population available (either in a zoo or in a confined habitat)? We still would never have the capability of knowing the true mean due to our inability to accurately measure. We can come close, very close to the truth but we can never reach it because our measurement tools are so inaccurate and crude. That is quite a statement indeed!

We confirm or **support** others when our data are similar to theirs and we **refute** (contest or challenge) a hypothesis when our data are statistically different. Our data and results are only just one small fraction of true, they certainly do not prove anything. Please keep the word "prove" out of your lab reports and lab exercise assignments, know what your data do or do not do to assist with the goals of scientific research.

Experimental Arrangement Examples Showing Significance

Now that you have learned about developing an experiment, collecting data, creating data sheets, experimental designs, and statistical significance, it is time to put them all together to make them work for you. Let us look at several examples of each of the experimental designs and statistical arrangements to determine if there is significance in the data. We are going to look at the CRD **statistical arrangement, a 1-way factorial,** which is how the treatments will be statistically related to each other. People get statistical arrangements and experimental designs confused all the time, but they are different. **Experimental designs** are how the experimental units are randomized in the field, pasture, feedlot, or nursery block. Statistical arrangements tell us if our data are significant or not. One-way factorial arrangements have one independent variable but it may be in different levels or rates.

Additionally, other statistical arrangements in agriscience experiments are provided at the end of this book in the Appendix. You will find an example Excel spreadsheet for each of the experimental designs discussed in Chapter 5 and a suggested statistical analysis from JMP® SE (Student Edition). That seems to be the biggest fault of experimental design and statistics courses, students are taught about different designs and then they are taught about different analysis techniques but they are not shown how to put them together.

CRD Analysis with Excel and JMP® SE software

We started our discussion of different experimental designs with the CRD and had several examples of them including height data collected from 30 weeping fig trees undergoing two fertilizer treatments (two different fertilizer brands, review Chapter 4). The data we got from them revealed their heights (cm) taken over several weeks, including initial and final heights.

In this section, we are going to analyze the data using two different statistical software programs, Excel and JMP® SE (Student Edition). Again, Excel is a well-known spreadsheet software package from the Microsoft® Corporation and is a component of Microsoft Office®. Excel is very robust and gives lots of options for data entry and analysis. Most students have access to Excel and other Microsoft® data management, presentation, and word processing software when they purchase their personal computer. Since Excel is found on most school computers and is readily accessible, it makes sense to start our tutorial with it.

JMP® SE is a statistics software package from the SAS Corporation. It is a student edition of JMP® that is an interactive program that links data on spreadsheets to statistical analyses and graphics displays. It has a greater capability than Excel does but that is to be expected: it was designed as statistical software. JMP® SE is readily available to those in academia though JMP® or the SAS Corporation and is recommended for professional scientists, engineers, analysts, and researchers. An advantage of using JMP® SE is its ability to quickly provide graphics at the touch of a button and additional statistical analyses at the click of a mouse. Excel is a good starter software but for advanced statistical analyses, seriously consider using JMP® SE. Its low cost and availability make it a good alternative when analyzing data.

Figure 6-2 shows the complete height data that we were first introduced to in Chapter 4 from the fig tree fertilizer experiment. What do these data tell us? We already learned how to calculate the means, create a graph, and even include trendlines in Chapter 4. But what we really want to know is: is there a difference between treatments? Does one fertilizer enhance growth more than the other? For this, we need to complete a t-test two times. The first t-test will tell us if our initial data were significantly different (we hope they are not, we want a homogeneous group to start with) and the second t-test will compare the final heights and tell if there was a significant difference between the treatments at the end of the study.

Microsoft Excel - making a data sheet with Excel 2003.xls

File Edit View Insert Format Tools Data Window Help Adobe PDF

I16

	A	B	C	D	E	F	G
2							
3	fertilizer	tree	initial ht	week 1	week 2	week 3	final ht
4	1	1	5.4	5.7	6.5	7.6	9.0
5	1	2	5.6	6.1	6.9	8	10.2
6	1	3	5.8	6.5	7.3	8.4	10.6
7	1	4	5.5	5.9	6.7	7.8	10
8	1	5	5.2	5.9	6.7	7.8	10
9	1	6	5.3	5.8	6.6	7.7	9.9
10	1	7	5.2	6	6.8	7.9	10.1
11	1	8	5.6	6.2	7	8.1	10.3
12	1	9	5.8	6.4	7.2	8.3	10.5
13	1	10	5.5	6.1	6.9	8	10.2
14	1	11	5.5	6	6.8	7.9	10.1
15	1	12	5.4	5.9	6.7	7.8	10
16	1	13	5.7	6.3	7.1	8.2	10.4
17	1	14	5.9	6.6	7.4	8.5	10.7
18	1	15	5.8	6.7	7.5	8.6	10.8
19	2	1	5.3	5.8	6.6	7.7	10.5
20	2	2	5.6	6.3	7.1	8.2	11
21	2	3	5.5	6	6.8	7.9	10.7
22	2	4	5.4	6	6.8	7.9	10.7
23	2	5	5.7	6.7	7.5	8.6	11.4
24	2	6	5.1	5.9	6.7	7.8	10.6
25	2	7	5.8	6.5	7.3	8.4	11.2
26	2	8	5.5	6	6.8	7.9	10.7
27	2	9	5.2	6.3	7.1	8.2	11
28	2	10	5.3	5.8	6.6	7.7	10.5
29	2	11	5.6	6.1	6.9	8	10.8
30	2	12	5.5	6	6.8	7.9	10.7
31	2	13	5.7	6.2	7	8.1	10.9
32	2	14	5.8	6.6	7.4	8.5	11.3
33	2	15	5.6	6.2	7	8.1	10.9
34							

p-value final / Sheet1 / ANOVA HW example / homework example

Figure 6-2. The complete sample height data (cm) collected from two fertilizers.

You can easily perform a t-test using Excel software by following several simple menu commands. Start off by having your Excel data in columns side-by-side with the labels at the top of each column. Go to the main menu for Tools (Figure 6-3), then drop down to Data Analysis and click on it. Most Excel software packages will have the Data Analysis option installed. If the Data Analysis command is not available, you need to load the Analysis ToolPak add-in program which is found on the original CD for Microsoft Excel.

The Data Analysis dialog box will appear and gives you several options to analyze your data (Figure 6-4). In Excel, you will not find the t-test listed but instead have the option of choosing the **Anova: Single Factor option.** This performs the same procedures that a t-test does; it is just named differently in Excel software. To continue with the analysis, select the Anova: Single Factor option and click OK. Excel is very user friendly and once you get the hang of the menus and steps, analyzing two different treatments will be easy for you each time you open the dialog box.

The Anova: Single Factor input dialog box will appear (Figure 6-5) requesting the range of values for the analysis. This step is similar to how we entered the range of data to make the graph back in Chapter 4. Remember how we highlighted the columns we wanted to include? Also on this dialog box there is the option of selecting the alpha value (you will see in Figure 6-5 it has already been set to 0.05 by default). Make sure you include the labels from the data columns if you had them above your columns of data on Excel. Having the column names is handy when going through the treatment results after the analysis. Click on **OK** to run the analysis to calculate the p-value.

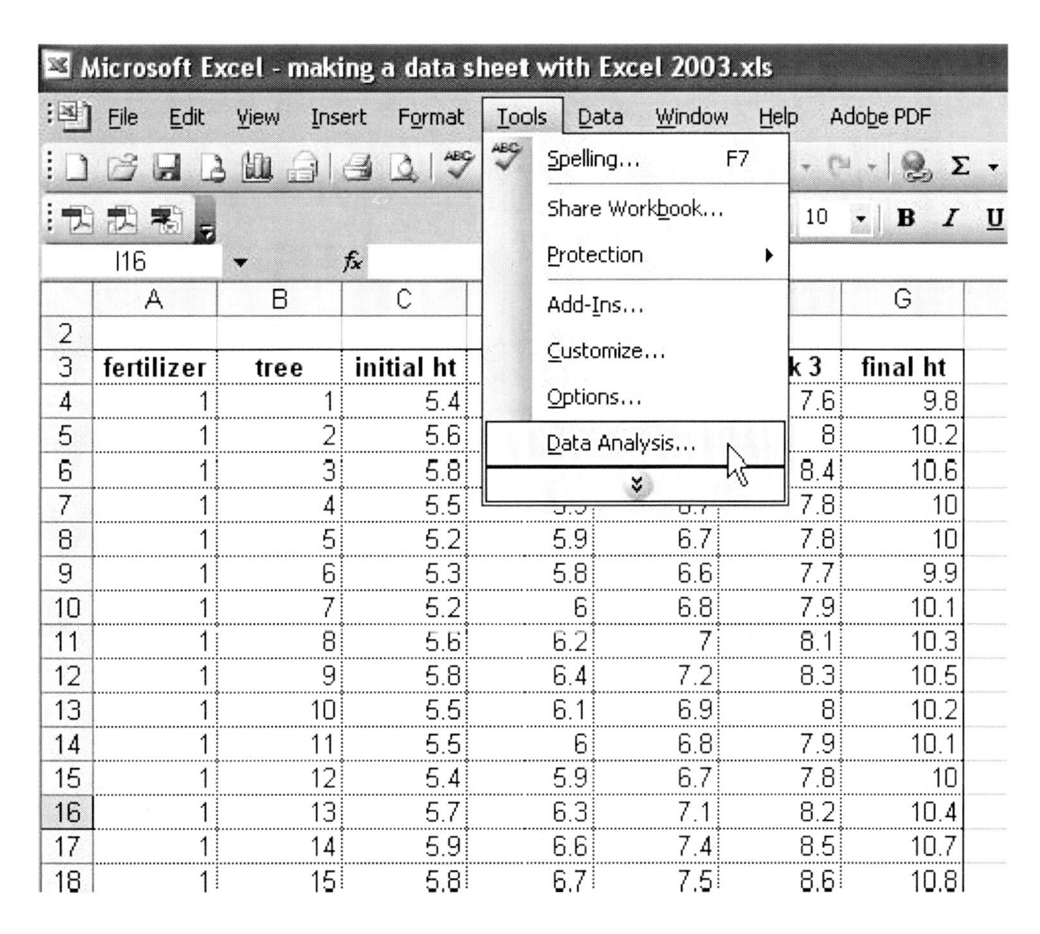

Figure 6-3. The first steps to completing a t-test or single factor ANOVA in Excel begins with the Tools menu where you can find the Data Analysis platform.

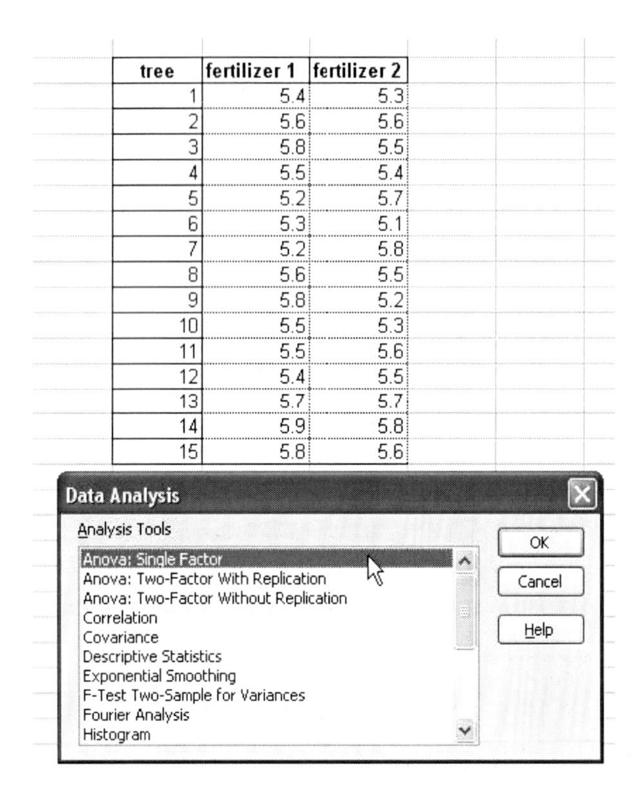

tree	fertilizer 1	fertilizer 2
1	5.4	5.3
2	5.6	5.6
3	5.8	5.5
4	5.5	5.4
5	5.2	5.7
6	5.3	5.1
7	5.2	5.8
8	5.6	5.5
9	5.8	5.2
10	5.5	5.3
11	5.5	5.6
12	5.4	5.5
13	5.7	5.7
14	5.9	5.8
15	5.8	5.6

Figure 6-4. Organize your initial height data into two columns on Excel and go through the Anova: Single Factor data analysis menu to determine significance between them.

tree	fertilizer 1	fertilizer 2
1	5.4	5.3
2	5.6	5.6
3	5.8	5.5
4	5.5	5.4
5	5.2	5.7
6	5.3	5.1
7	5.2	5.8
8	5.6	5.5
9	5.8	5.2
10	5.5	5.3
11	5.5	5.6
12	5.4	5.5
13	5.7	5.7
14	5.9	5.8
15	5.8	5.6

Anova: Single Factor
Input
Input Range: $J$3:$K$18
Grouped By: Columns / Rows
Labels in First Row
Alpha: 0.05
Output options
Output Range:
New Worksheet Ply:
New Workbook
OK
Cancel
Help

Figure 6-5. Organize your initial height data into two columns and go through the Anova: Single Factor data analysis menu to determine significance.

The Anova: Single Factor Summary table and ANOVA tables for the initial height data are given on a new worksheet (Figure 6-6) since the New Worksheet Ply Output option was checked. You can find the treatment means, data counts, F statistic, and most importantly, the p-value. The p-value for our initial data in the fertilizer experiment is 0.618721. Since this value is greater than our predetermined $\alpha = 0.05$, we do not have significance and therefore fail to reject the null hypothesis. There is no difference in our initial height treatment means. What does this mean exactly? This means that our initial heights were not statistically different so we conclude they were a homogeneous group of weeping figs to start our trial. Now, most people start to think wow, what if the treatment means were statistically (significantly) different at the beginning? That would be a real cause for concern because we would not know if our final data resulted from the treatment effects or because one group was larger to begin with.

I hope you see how important it is to take initial data immediately upon selecting experimental units and not wait too long in the experiment to check to see if there are significant differences among them. If there are initial differences (and in agriscience experiments it occurs quite often), blocking similar experimental units into groups can be a way to work around the variation and still utilize them in the study. So, we see there are no significant differences between our plants at the beginning of the experiment and that is a good thing! Blocking into homogeneous groups will be discussed later on.

Let us go through the final height data from this fertilizer experiment and see how they turned out. Was there a significant difference between the final heights? Use the same steps as seen with the initial heights to calculate the single factor ANOVA (t-test) on Excel. Go to the **Tools menu** and then click on **Data Analysis.** Locate the **Anova: Single Factor** menu again (Figure 6-7) and click on **OK.**

	A	B	C	D	E	F	G
1	Anova: Single Factor						
2							
3	SUMMARY						
4	*Groups*	*Count*	*Sum*	*Average*	*Variance*		
5	treatment 1	15	83.2	5.546667	0.04981		
6	treatment 2	15	82.6	5.506667	0.044952		
7							
8							
9	ANOVA						
10	*Source of Variation*	*SS*	*df*	*MS*	*F*	*P-value*	*F crit*
11	Between Groups	0.012	1	0.012	0.253266	0.618721	4.195972
12	Within Groups	1.326667	28	0.047381			
13							
14	Total	1.338667	29				
15							

Figure 6-6. The Anova: Single Factor data summary provides sample means of each treatment and the p-value.

The Anova: Single Factor input box appears again (Figure 6-8), confirm the alpha is 0.05 and that the labels will appear in the output (since the labels are included in this example as headings above the final heights, we need to click on Labels in First Row in the input box).

As seen before with the initial height data, the Anova: Single Factor Summary table and ANOVA tables for the final height data will be provided on a new worksheet (Figure 6-9) since the New Worksheet Ply Output option was checked.

The Anova: Single Factor SUMMARY sheet gives the treatment means, data counts, and p-value. The p-value for our final height data in the fertilizer experiment is small at 2.68 x 10^{-6} (Figure 6-9). Since this value is less than our $\alpha = 0.05$ (our p-value is 0.00000268), we do have significance and therefore reject the null hypothesis. There is a difference between our treatment means, with fertilizer 1 having a height average of 10.24 cm and fertilizer 2 having a height average of 10.86 cm. We can conclude that fertilizer 2 contributed to the height gain in weeping fig plants and would probably be a good fertilizer to use again. Would we totally disregard fertilizer 1 and not use it at all? Probably not; it did work, but fertilizer 2 worked better in this case.

Have you thought about using JMP® SE statistical software to determine significance in data? There are several very good computer programs on the market and JMP® SE is very user friendly. You have the option of working directly in the software or importing your data from another source, such as Excel. We will use our weeping fig height data from our fertilizer example experiment to show you that it is quite simple to use and very powerful to find answers quickly. There are several things that can be found out easily like we saw in Excel, but having access to graphics and other features to tell you more information about your data will be demonstrated to you through this next tutorial.

To use JMP® SE, open the program and you will see that you have several choices of files to open. Click on **Open Data Table** button (Figure 6-10) on the JMP Starter menu. It will ask you to select either an existing JMP file (Figure 6-11) or an existing file from Excel or another spreadsheet program (Figure 6-12). Since you probably will not have any JMP files made yet, click on the Open Data Table button and search your computer for the spreadsheet file that has the data sheet you wish to open or import. Your instructor may put sample Excel files on the class website, ask if these are available for you to download. If the file you need to open is a JMP file, simply open it, if it is another type of file, you might need to go to **All Files** under **Files of Type** on the open dialog menu (Figure 6-12) for it to appear.

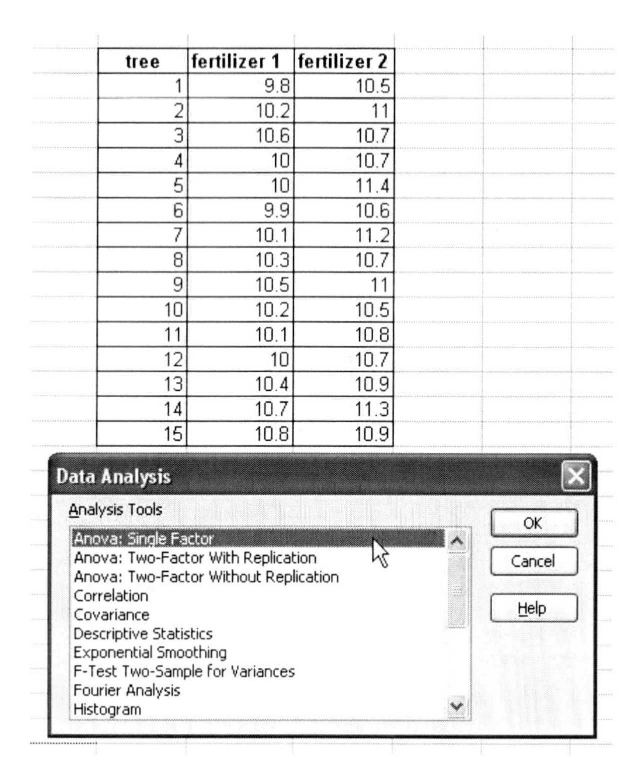

Figure 6-7. Organize your final data into two columns and go through the Anova: Single Factor data analysis procedure to determine significance among treatments.

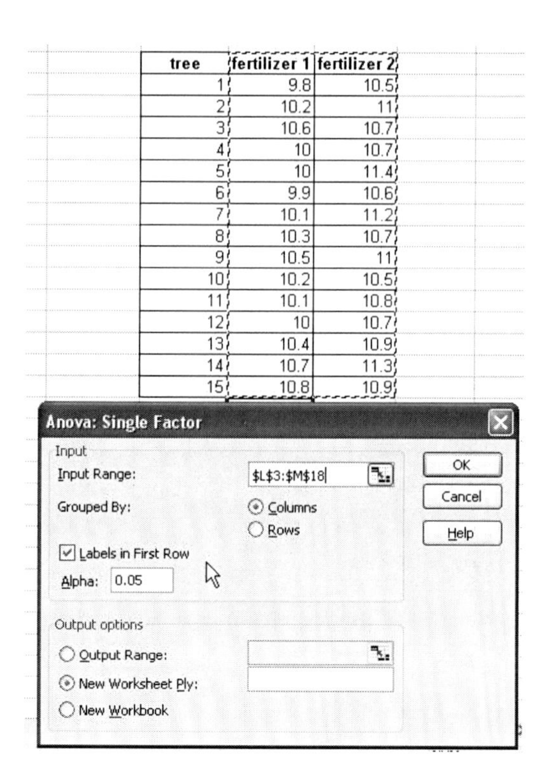

Figure 6-8. The final height data is highlighted into two columns and entered into the Anova: Single Factor data analysis menu to determine significance.

	A	B	C	D	E	F	G
1	Anova: Single Factor						
2							
3	SUMMARY						
4	*Groups*	*Count*	*Sum*	*Average*	*Variance*		
5	fertilizer 1	15	153.6	10.24	0.091143		
6	fertilizer 2	15	162.9	10.86	0.076857		
7							
8							
9	ANOVA						
10	*rce of Varia*	*SS*	*df*	*MS*	*F*	*P-value*	*F crit*
11	Between G	2.883	1	2.883	34.32143	2.68E-06	4.195972
12	Within Gro	2.352	28	0.084			
13							
14	Total	5.235	29				
15							

Figure 6-9. The Anova: Single Factor final fertilizer treatment height data summary provides sample means of each treatment and the p-value.

Figure 6-12 shows both an Excel file (JMP SE data.xls) and a JMP file (HOH data.JMP), after the menu **All Files** was selected. In this tutorial, the JMP HOH data file was selected for this example and has the same data as the weeping fig fertilizer data we used in the Excel tutorial previously.

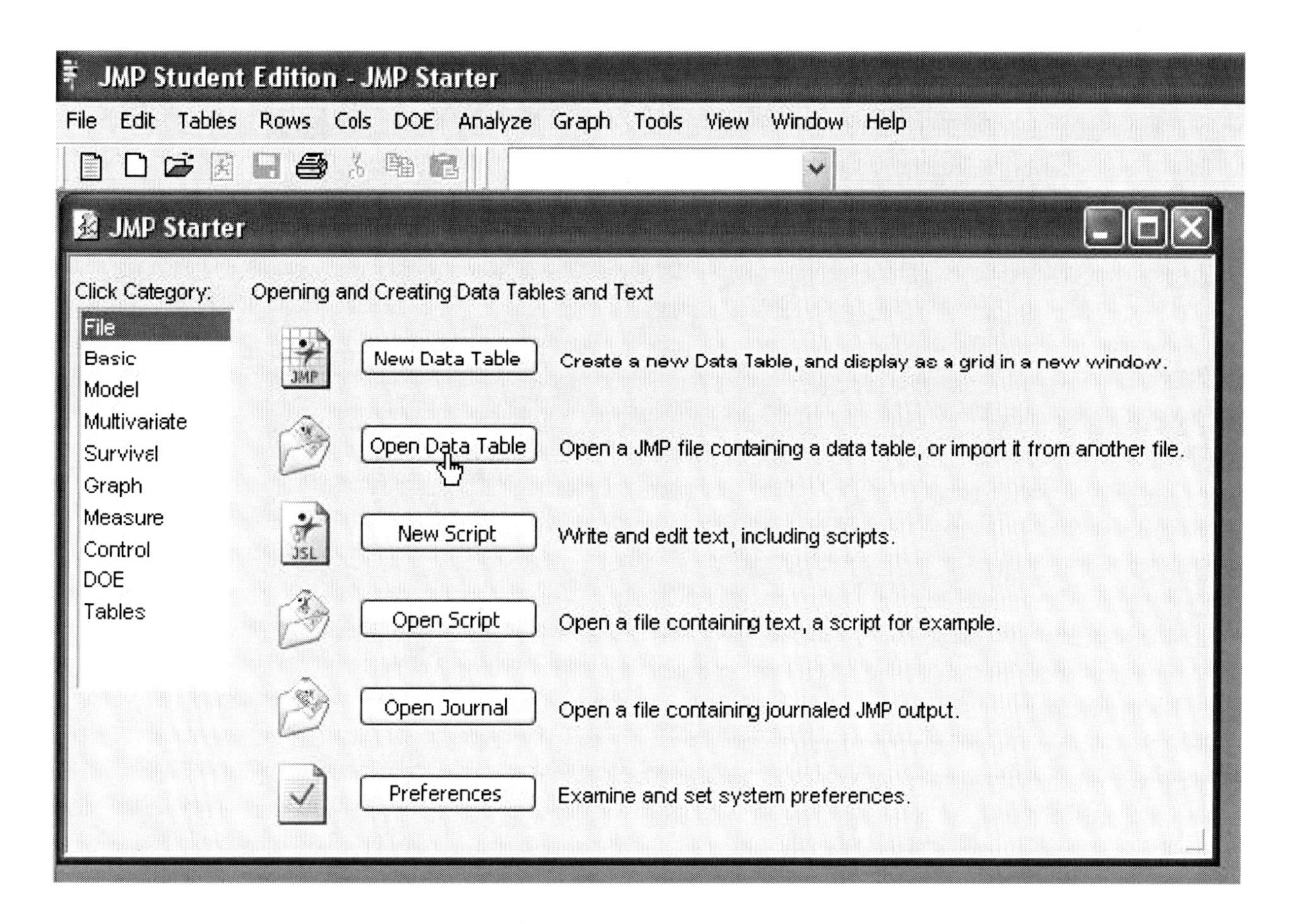

Figure 6-10. Opening an existing file in JMP® SE can be done with the Open Data Table button.

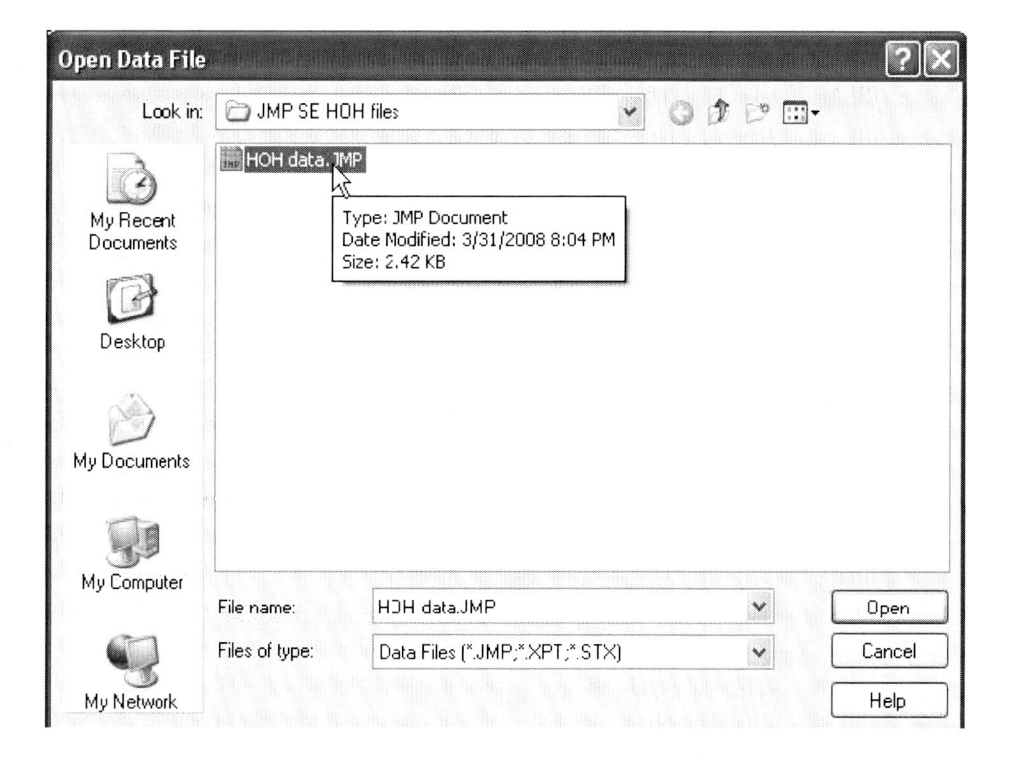

Figure 6-11. Bringing an existing file into JMP® SE.

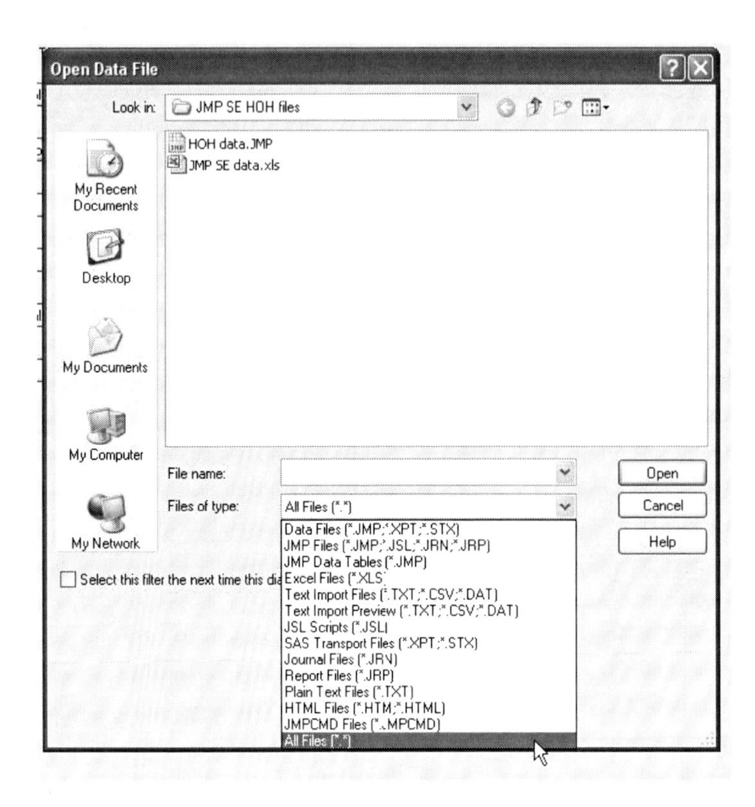

Figure 6-12. Finding existing files in JMP® SE under the All Files (*.*) button.

As you look at this new JMP data table (Figure 6-13), note how it looks compared to what you have already seen in Excel. It has rows and columns, your data are in the same columns, there are headers at the top of each column, and the menus across the top should look similar to some other spreadsheet or word-processing programs you have previously used.

The JMP columns all come as continuous when they are imported and that is great if you want numerical data to work with. The data in our columns come in both as numerical data and as coding variables. Remember back in Chapter 3 when we made coding variables to make our labels for our potted plants? In Table 3-1, we assigned each treatment to a coding variable (control had the coding variable number 1, the lowest fertilizer concentration treatment had coding variable number 2, etc.). This allowed us to name our replicates on the label with simple codes, eliminating a lot of writing on each label, a real timesaver!

Plant 2-1 was the first replicate from treatment 2, plant 4-5 was the fifth replicate from treatment 4, and so on. This is no different than what we see in our imported data table, the difference is now we have to tell JMP that certain columns are not numbers in the sense of being data values but numbers in the sense of being coding variables, they take the place of names.

Telling JMP how to change column information is as easy as a click of the right mouse. Right mouse click on the top of a column, such as seen in Figure 6-14. A menu appears asking you to choose what you want to do, click on **Column Info**.

For the fertilizer column (Figure 6-15), you will see the column name (leave it as it is), the data type (it is a number so leave it as it is), and modeling type. Leave the format and column properties buttons alone for now, focus back on the modeling type. There are three modeling types you can choose:

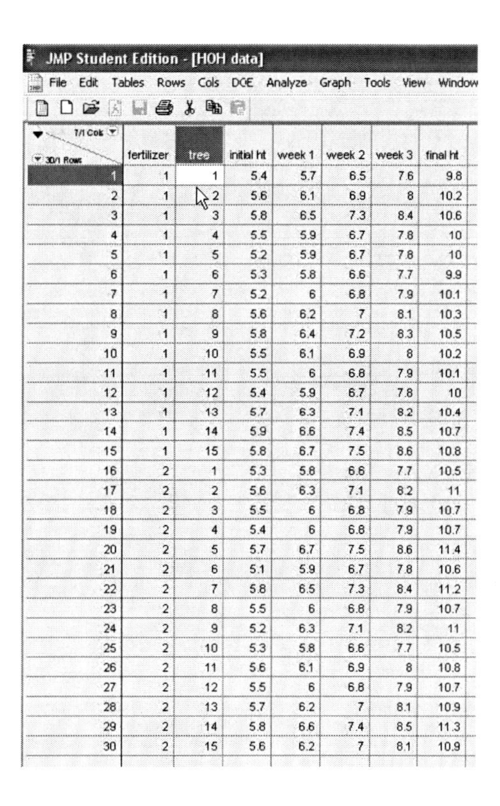

JMP Student Edition - [HOH data]

	fertilizer	tree	initial ht	week 1	week 2	week 3	final ht
1	1	1	5.4	5.7	6.5	7.6	9.8
2	1	2	5.6	6.1	6.9	8	10.2
3	1	3	5.8	6.5	7.3	8.4	10.6
4	1	4	5.5	5.9	6.7	7.8	10
5	1	5	5.2	5.9	6.7	7.8	10
6	1	6	5.3	5.8	6.6	7.7	9.9
7	1	7	5.2	6	6.8	7.9	10.1
8	1	8	5.6	6.2	7	8.1	10.3
9	1	9	5.8	6.4	7.2	8.3	10.5
10	1	10	5.5	6.1	6.9	8	10.2
11	1	11	5.5	6	6.8	7.9	10.1
12	1	12	5.4	5.9	6.7	7.8	10
13	1	13	5.7	6.3	7.1	8.2	10.4
14	1	14	5.9	6.6	7.4	8.5	10.7
15	1	15	5.8	6.7	7.5	8.6	10.8
16	2	1	5.3	5.8	6.6	7.7	10.5
17	2	2	5.6	6.3	7.1	8.2	11
18	2	3	5.5	6	6.8	7.9	10.7
19	2	4	5.4	6	6.8	7.9	10.7
20	2	5	5.7	6.7	7.5	8.6	11.4
21	2	6	5.1	5.9	6.7	7.8	10.6
22	2	7	5.8	6.5	7.3	8.4	11.2
23	2	8	5.5	6	6.8	7.9	10.7
24	2	9	5.2	6.3	7.1	8.2	11
25	2	10	5.3	5.8	6.6	7.7	10.5
26	2	11	5.6	6.1	6.9	8	10.8
27	2	12	5.5	6	6.8	7.9	10.7
28	2	13	5.7	6.2	7	8.1	10.9
29	2	14	5.8	6.6	7.4	8.5	11.3
30	2	15	5.6	6.2	7	8.1	10.9

Figure 6-13. An imported file in JMP® SE appears in the same format as the original spreadsheet.

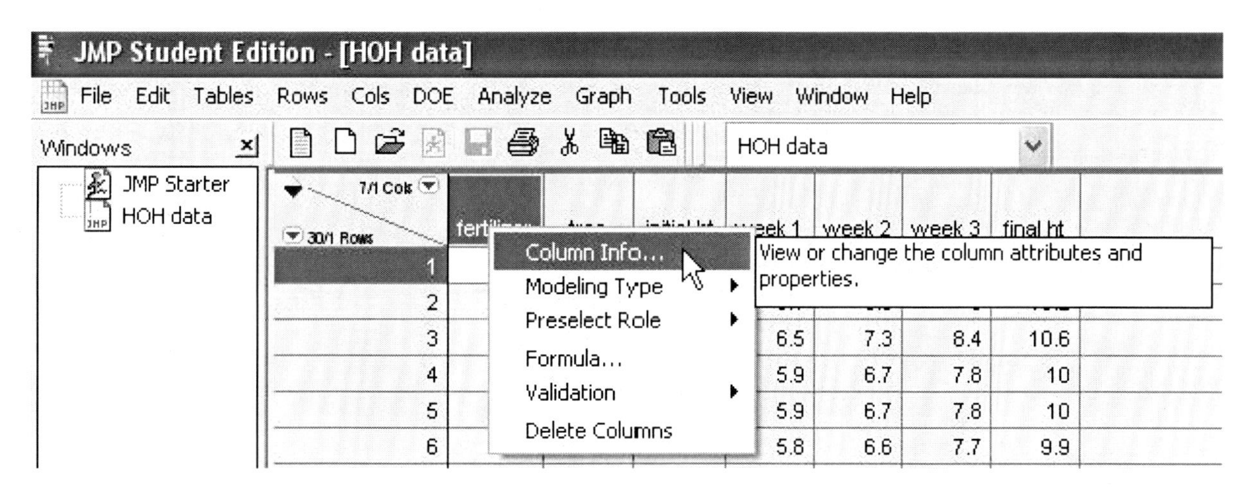

Figure 6-14. Existing column information can be changed at each column header.

1. Continuous – these are numbers measured on a continuous scale or range, they typically include numerical data and are values
2. Ordinal – these are numbers that represent an order or ranking such as "high," "medium," and "low" or "large" or "small"
3. Nominal – these numbers or other variables (which can include names and words) are the categories that the data represent

The modeling types of the variables determine which analyses can be performed. Since the heights are examples of continuous data, Figure 6-16 shows what the continuous modeling type looks like for height and it is most likely already in that format.

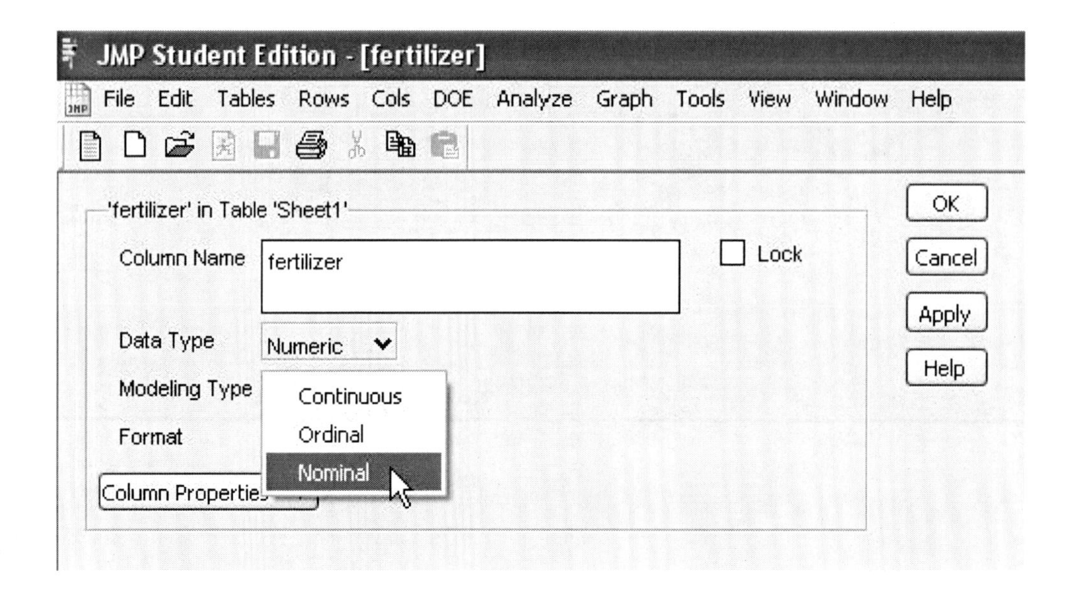

Figure 6-15. Completing column name changes or modeling type changes to Nominal.

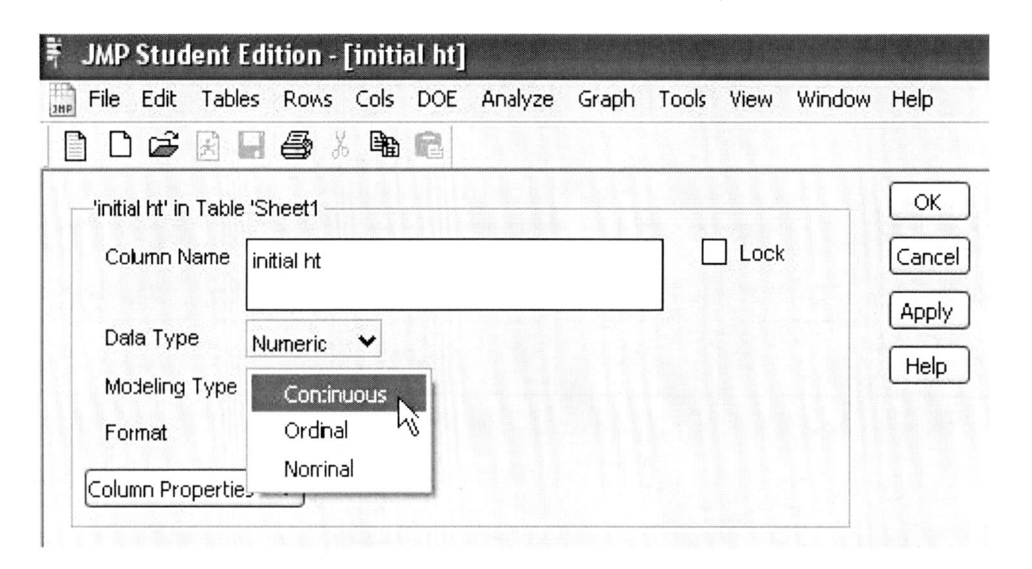

Figure 6-16. Completing column name changes or modeling type changes to Continuous.

When you have your data in the correct modeling type, it is time to run our analysis. We are going to do the same analysis that we did with Excel earlier and compare the results; but you should also know that JMP® SE provides summary statistics using the **Distribution** menu. One of the advantages of using the JMP software is the graphics that come with many analyses, which makes it easier to "read" the data quickly and find what you want to know.

One of the graphics available to you is called a **histogram**. A histogram is a bar graph that represents the frequency distribution. Histograms are helpful when looking at data for the first time because they can quickly show trends (remember the trendline from the Excel tutorial?). JMP not only makes histograms under the Distribution menu, it also gives simple descriptive statistics including the **count frequencies, mean, median, quartiles** (the 25th, 50th, or 75th percentiles of a frequency distribution divided into four parts, each containing a quarter of the population), and **standard deviation** (the measure of the data frequency distribution and dispersion around the mean). Descriptive statistics are very commonly presented from research projects.

To make a histogram with JMP and determine other descriptive statistics about your data, go to the **Analyze** menu to find **Distribution** and select it (Figure 6-17).

From there, you will be taken through a series of steps to add the columns of data you wish to explore through the Distribution platform. Let us look at our columns of interest, the fertilizer column and the final heights column. We select these two columns and add them to the **Y, Columns** selection box (Figure 6-18) because we want to see what the data by fertilizer and final heights are. Click on OK. Again, this is a quick way to look at the data and see what it can initially tell us before we move on to the next step.

The Distributions dialog box appears with the histograms for both fertilizer type (1 and 2) and for the final heights (in cm) in Figure 6-19. There is a lot of information in this box so let us take a moment and go through each part. Clicking on either of the fertilizer 1 or 2 boxes in the histogram on the left highlights the corresponding data in the histogram on the right. Figure 6-19 shows the highlighted final height data for the first fertilizer treatment and that the lowest final heights came from that treatment.

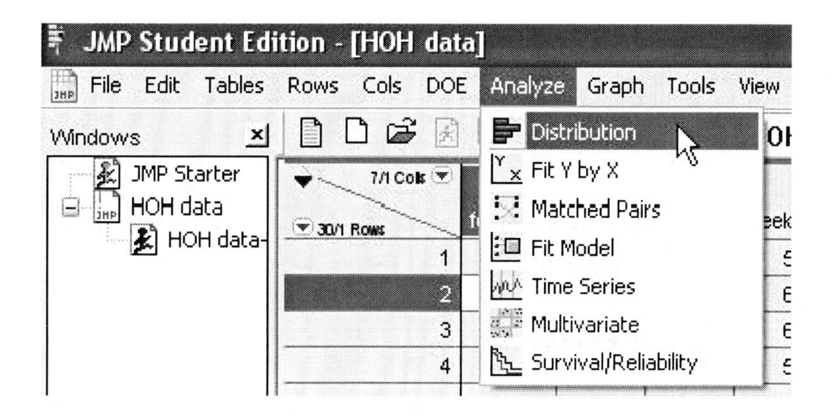

Figure 6-17. Creating a histogram starts with the Distribution platform.

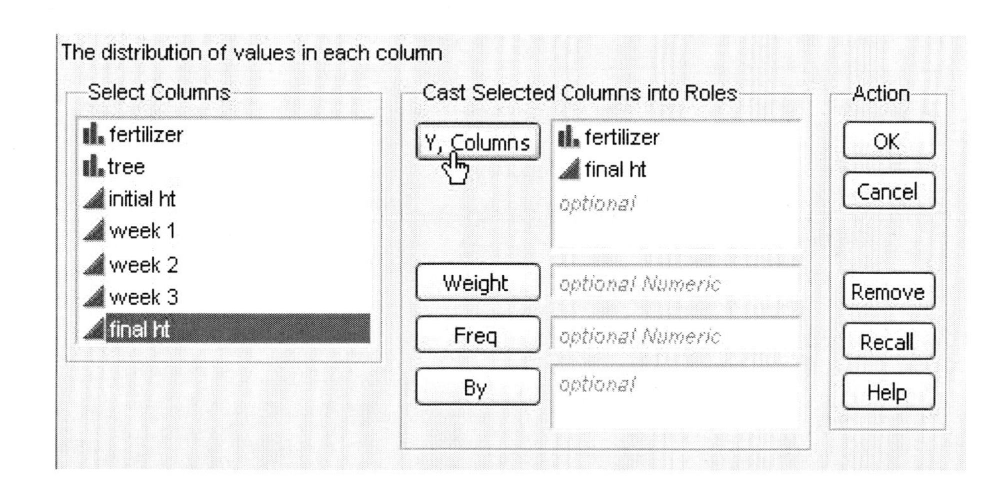

Figure 6-18. Select the data columns you wish to include in the histogram.

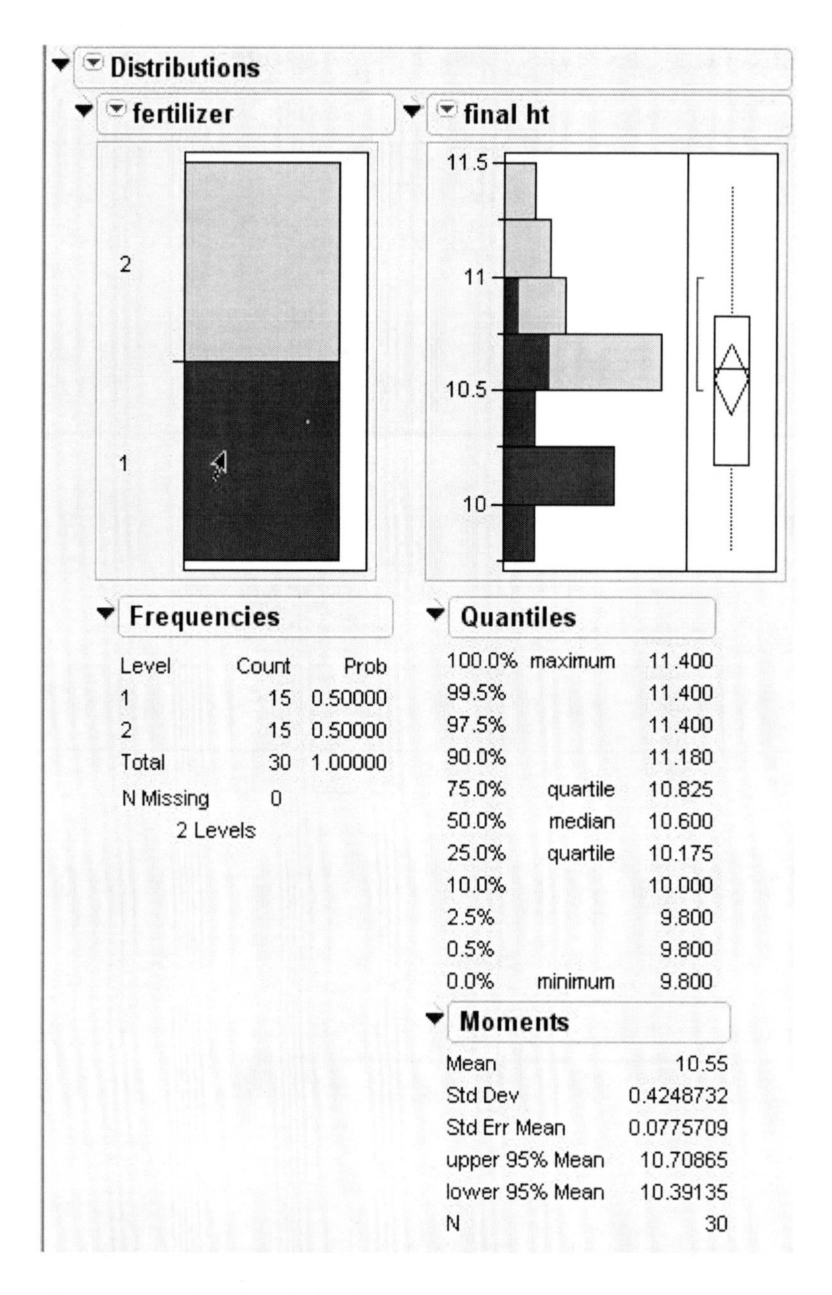

Figure 6-19. Select the data columns you wish to view in the final height histogram.

If looking at the histograms with their bars along the side is confusing, you can easily switch the histogram and have it turned with the bars rising up (Figure 6-20). To do this, click on the **red triangle** (which appears gray in this text) next to the fertilizer header above the histogram, scroll down to **Histogram Options**, and toggle the **Vertical** option on or off. Repeat with the **final heights red triangle**.

As you click between the fertilizers, you can see the data on the second histogram become highlighted as well. The frequency data do not change but the graphic can be a powerful tool when looking at data for the first time. Click on each type of fertilizer and you will see that fertilizer turn dark. The corresponding heights will turn that dark color as well, showing you which heights came from each treatment.

When you are happy with the initial viewing of the data from the Distribution menu and want to analyze your data with a t-test or an ANOVA, it is time to go back to the main menu heading (you can leave the Distribution box by closing it with the x button in the upper right-hand corner of the

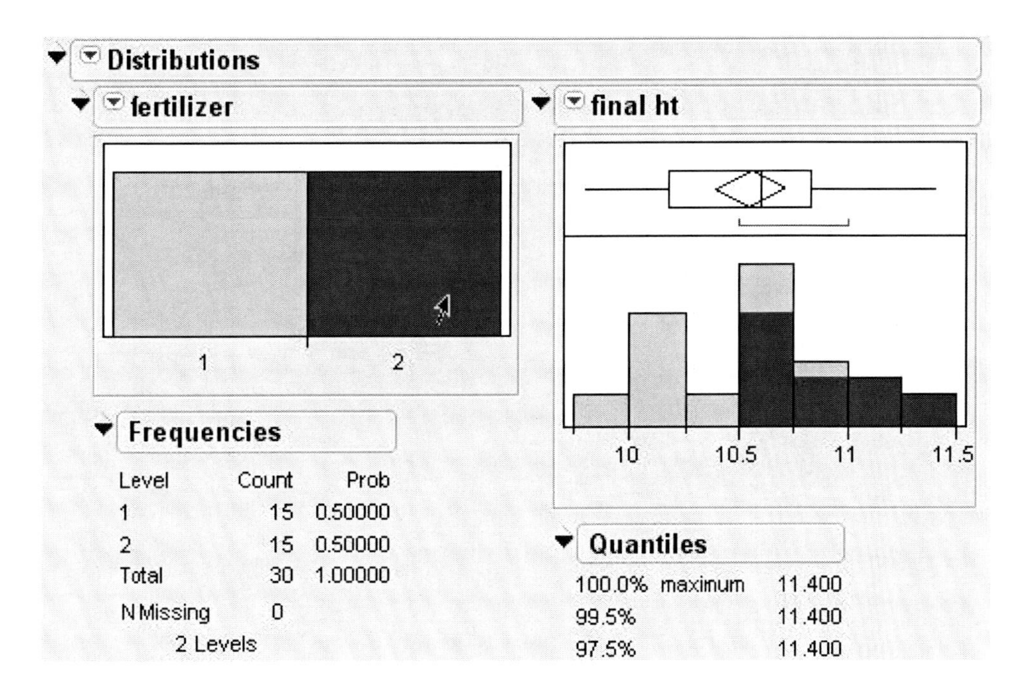

Figure 6-20. The histograms can be turned horizontally for better viewing.

screen). Once the Distribution box is closed, you should be back on the main data JMP spreadsheet page. Now it is time to see if there is a significant difference between our treatments. To get started, select **Analyze** at the top of the screen, then select **Fit Model** (Figure 6-21). We are going to see if our data will fit into the model for the null hypothesis that there are no treatment differences, and that is the model we are creating with our data.

By the way, a **model** in scientific research is a mathematical representation of a simplified explanation. When we calculate some statistical test, we are trying to find out if the data from our experiment can be explained by statistics. That is what a p-value tells us; it is the probability (the "p" in p-value) that the test statistic would be calculated if the null hypothesis is true. So a p-value of 0.02 is the probability that the null hypothesis is true. That is 2/100 and it is a very small chance that the null hypothesis (there are no differences between the treatments) is true. Therefore, we reject the null hypothesis and state that our alternative hypothesis is correct. Remember that we want our p-value to be significant and that occurs when it is calculated less than 0.05, the alpha value that we agreed on ahead of time.

Let us go through the steps here on JMP for the initial height data and compare them to our Excel results to determine if either software program can be used to determine significance.

In the **Model Specification** box, you will need to put the independent variable (the treatment heading, what is being tested) fertilizer in the **Construct Model Effects** box (Figure 6-22). You will see that fertilizer column was selected under **Select Columns** on the left-hand side and then the **Add** button was selected. This places the fertilizer independent variable in the correct place (Figure 6-23).

To place the dependent variable to be tested (initial height), highlight initial height (Figure 6-23) and select the Y box in the **Pick Role Variables** box. This sets up the model that we are going to test the different fertilizers to see if their initial heights are statistically the same (that there is no different between them, as the null hypothesis would dictate). Click on Run Model to run the analysis.

The **Whole Model** screen appears (Figure 6-24), giving the reader many things to observe. With a quick glance, the p-value can be found in the middle of the left-hand column, and there are options

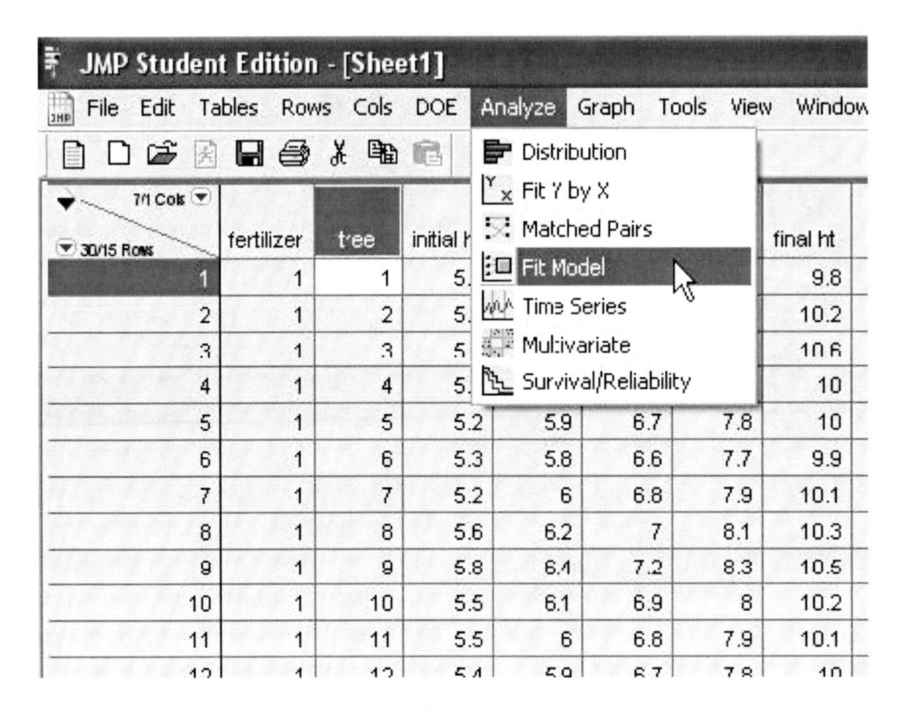

Figure 6-21. Running the analysis on the weeping fig fertilizer data starts with Fit Model.

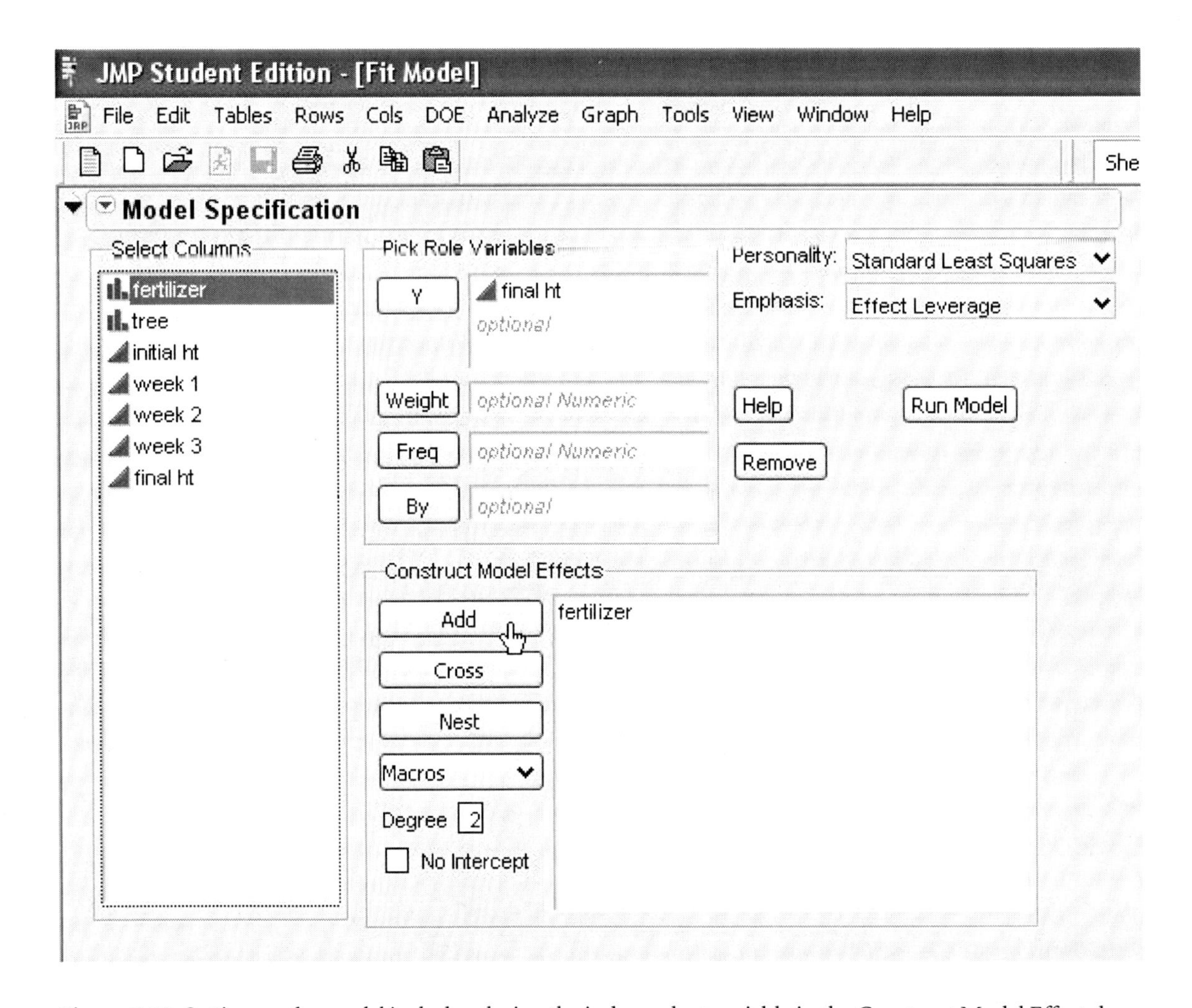

Figure 6-22. Setting up the model includes placing the independent variable in the Construct Model Effects box.

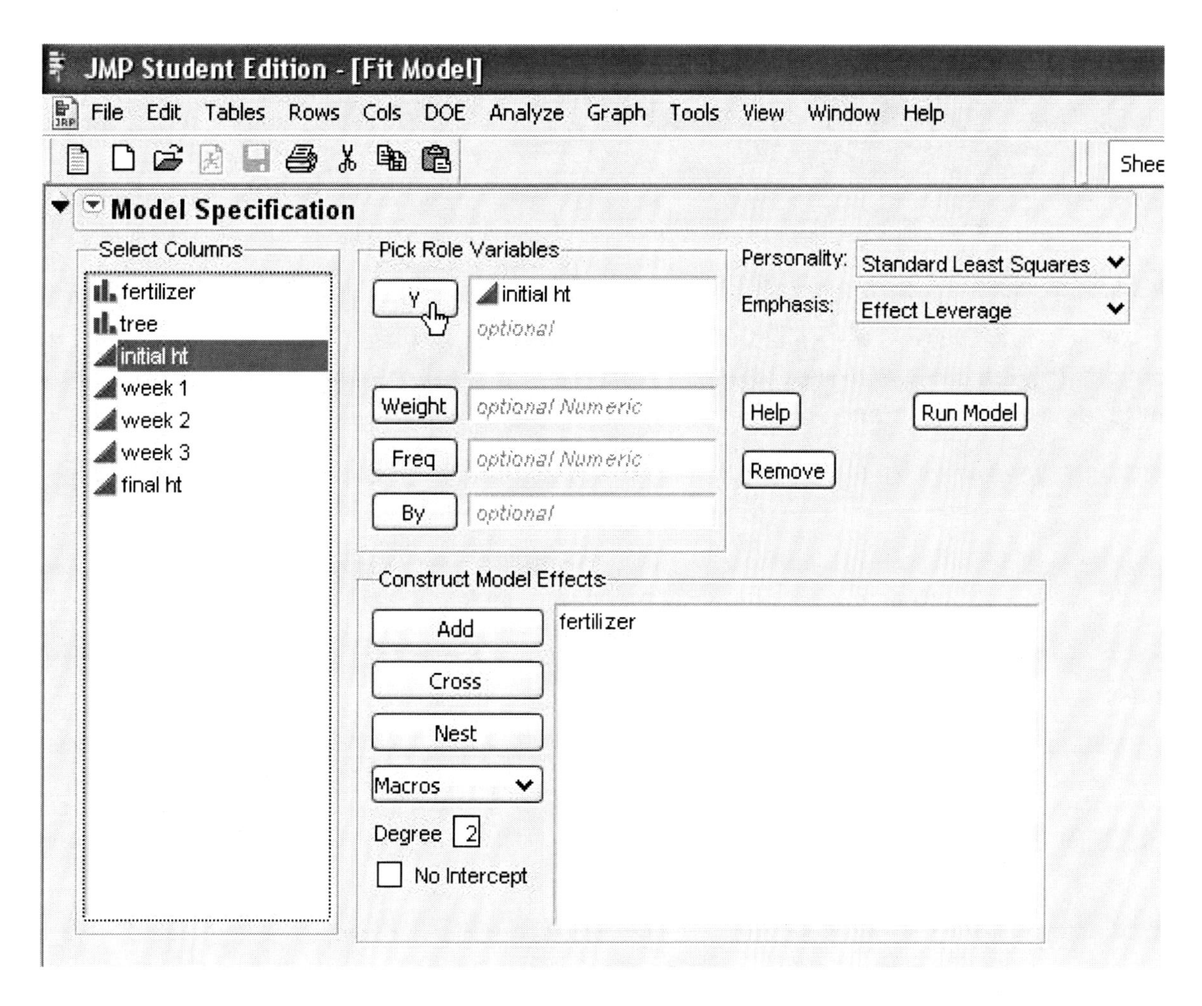

Figure 6-23. Setting up the model concludes by placing the dependent variable in the Pick Role Variables Y box. Initial heights were what we measured.

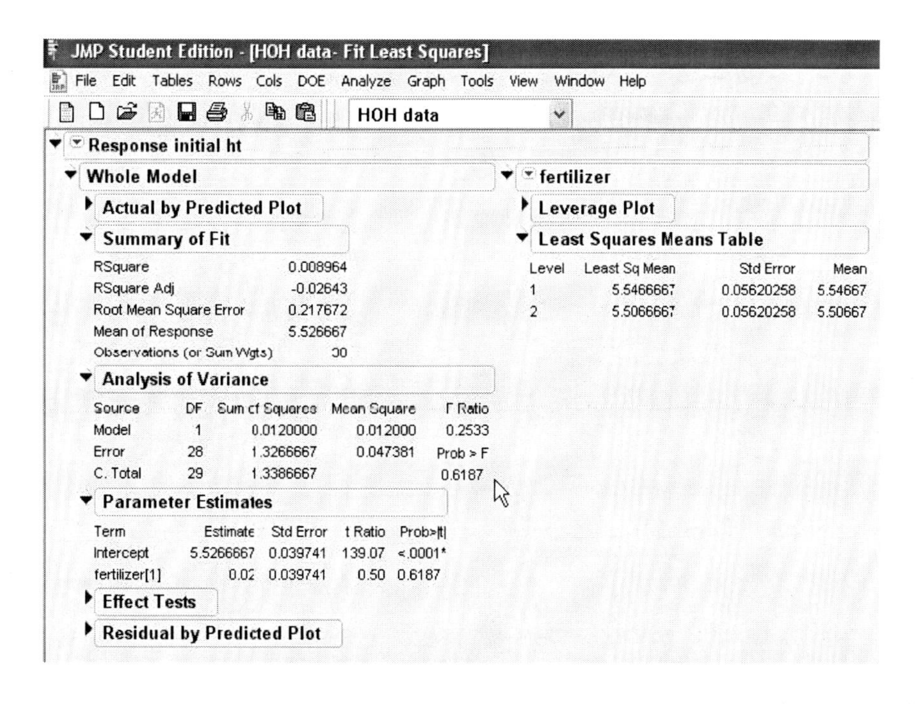

Figure 6-24. The Whole Model results give the initial heights p-value and ANOVA information calculated from the experiment.

for you to look at the data graphically. The right-hand column will provide specific information about the treatments in the experiment, as noted by the "fertilizer" heading with the red triangle (Figure 6-25).

Going back to the left-hand column, you can find the p-value of 0.6187 (by the large arrow) in Figure 6-26. This value is greater than our predetermined alpha of 0.05 so that means our data are not statistically significant. This is the probability that our data would appear due to random chance if the null hypothesis were correct. So, we can take that as being either our initial height data from both treatments really are statistically the same, or we got our results of them being the same due to random chance. Either way, we should be glad they are statistically the same since we normally want our initial data to be the same to eliminate bias between treatments.

fertilizer

Figure 6-25. Additional analysis menus are found on JMP with pull-down tabs like the triangle found next to the fertilizer heading.

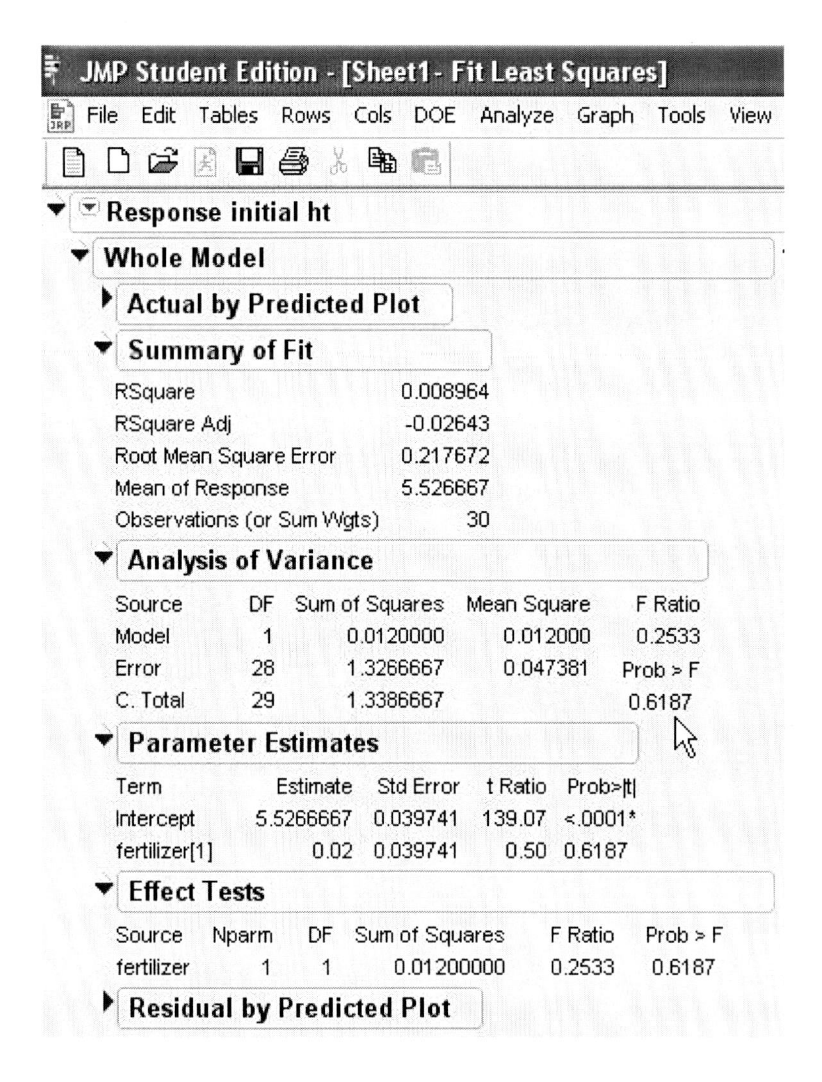

Figure 6-26. The Whole Model results give the ANOVA table that includes the p-value (next to the large arrow on the right) for the initial fertilizer data.

We can use the information provided in the right-hand column to tell us more information (and create a graphic) from the fertilizers. Clicking on the small triangle next to the fertilizer heading in the upper right-hand column (as previously seen in Figure 6-25), we find a pull down menu that lists several options. One option is for LS Means Plot (Figure 6-27).

This plot allows JMP to create a graph for you (Figure 6-28) that shows a comparison between the initial mean heights of the two fertilizer treatments. This is a quick way to see differences in physical form, which is an easier way for some people to view data.

Not everyone can look at numbers and see similarities or differences quickly, but having the option of looking at data in graphical form can help those of you who like more concrete examples. Figure 6-28 shows almost a flat line between the treatments, the treatment means must be very similar. Can you see that fertilizer 1's initial plant heights were a bit larger than those of fertilizer 2? Can we tell from the graph if they were statistically different? We cannot; and that is why we have to refer back to the p-value to give us that information. We know they are not significant since the p-value was 0.6187, which is greater than the $\alpha = 0.05$.

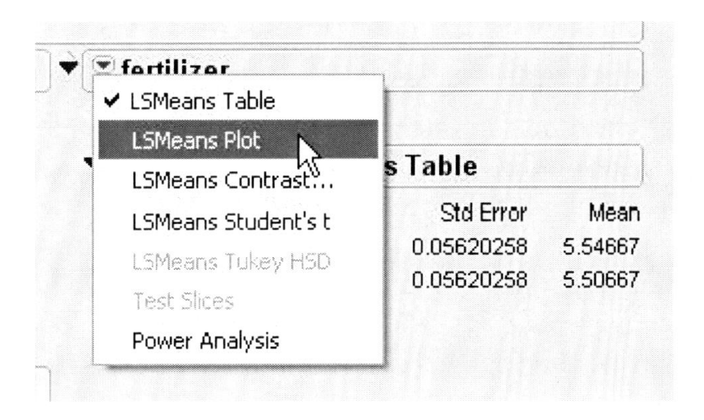

Figure 6-27. Creating an LS Means plot is started by scrolling down the menu under the inverted red triangle.

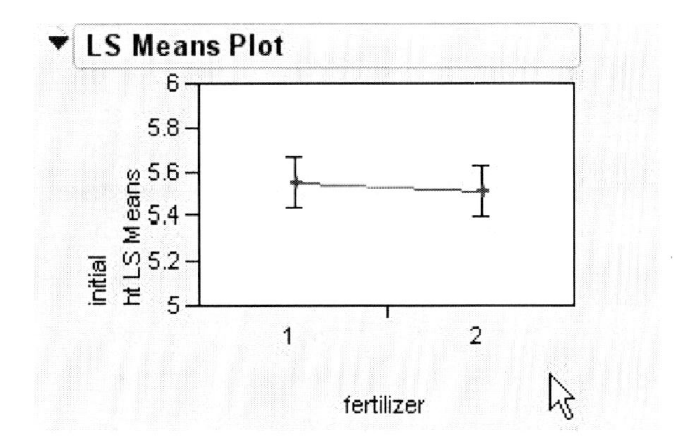

Figure 6-28. The LS Means plot for the initial height data shows the total mean value for each fertilizer treatment.

Using the Tukey Multiple Comparisons Analysis to Determine if Data are Statistically the Same as Other Data Values

You know that the standard grading scale for assignment and course grades in most classes is the following:

A = 90 – 100%

B = 80 – 89%

C = 70 – 79%

D = 60 – 69%

F = 59% and below

At the end of the semester no matter if you earn an 83% or an 88%, you will earn a B grade in the course. Both the 83% and the 88% are statistically the same, both are worth a B. Likewise, a 70% and a 71% are the same, a 94% and a 99% are the same, and so on. You already know about means comparisons, that comparing results can determine if they are similar or not, and you know this since you know how grades work, that is, you know how they are determined.

Have you ever had a class where the grades were so low the instructor had to curve them? What if there was an exam and the high exam score was 80/100. That student earning the 80 points would probably get the A grade and the others would be placed in the B, C, D and F grade categories accordingly. Perhaps there were 5 scores in the 70s (who all earned Bs) because of the curved grades: 71, 75, 76, and 2 had earned 78. It would not matter that someone earned a 71 and someone else earned a 78, both are Bs, right? Right, they are equal as far as the letter grade is concerned. How do professors do this? Some use Excel to compute the point totals, compute the means and standard deviations, and compare these against the class average. Others just rank the grades and see if there are natural breaks or groups in between the scores, and some take the grades strictly as earned with no curve added in until the end of the class and work with the final grades.

How does this discussion of curved grades help us understand if our height data are statistically the same or not? Well, what would we do if our initial height data from the fertilizer treatments had been significantly different with a p-value of less than 0.05? By only having two fertilizer options, we would know which one was the taller group and which was the shortest just by looking at the values.

But what if we had three or more treatments that we were comparing? We could run another analysis that could compare each treatment to each of the other treatments to see if they were similar treatments or not. This analysis of comparing means between several treatments is called **means comparisons** and one example of means comparisons is the **Tukey procedure**, named after Dr. John W. Tukey, a renowned statistician. JMP can run the Tukey means comparisons with just a couple of clicks of the mouse and gives the researcher another tool to interpret data. One thing to remember though is to only look at the Tukey means comparisons if you know from your p-value that your treatment results are statistically different, otherwise you know you are not going to find any treatment differences.

As an example from our initial height data, we know there are no significant differences between treatments because our p-value from the ANOVA analysis was 0.6187. Both fertilizer treatments' means are statistically the same since our p-value was greater than 0.05.

If you are curious and want to confirm that there is no significant difference, go to the fertilizer menu (red triangle) on the right-hand corner and pull down the menu to the **LSMeans Student's t** option (Figure 6-29). You will also see right below that an option for LSMeans Tukey HSD which is grayed out/not available. These will calculate the means comparisons test for you, which will put your data

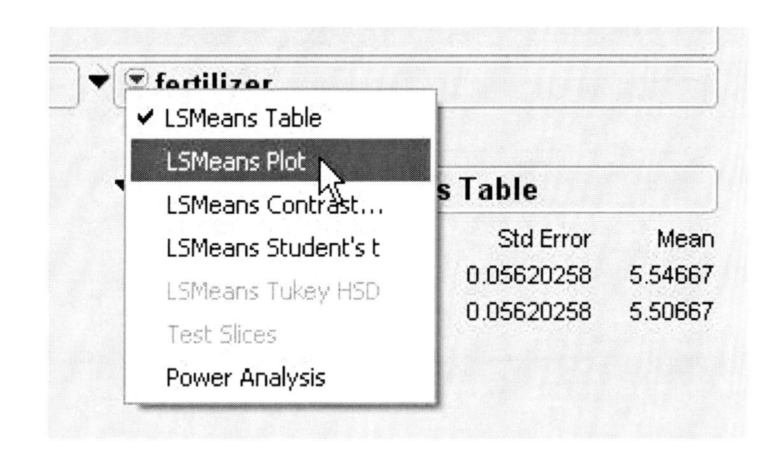

Figure 6-29. Calculating a means comparison for initial height begins with selecting either the LSMeans Student's t or LSMeans Tukey HSD.

into groups much like the exam grades we were just talking about. Since there are only two treatments in this experiment, the only option on JMP is the related Student's t, which is equal to running the **LSMeans Tukey HSD** analysis in experiments that have three treatments or more. Select the LSMeans Student's t option to see if there are treatment differences. Again remember we normally do not do this when there is no statistical difference but this is for practice and to introduce you to the concept of seeing significant differences by treatment groups.

When you click on this option, the **LSMeans Differences Student's t** box appears and shows the means comparisons between both groups; but, more importantly, it shows the statistical significance between groups labeled as capital letters. In Figure 6-30, both treatments (levels 1 and 2) have the level A and according to the statement found below them, levels not connected by same letter are significantly different. Since they both are assigned the letter A, they are statistically the same value; their heights are mathematically the same and could be interchanged. Remember our discussion of curved grades and that if you earn an 83% or an 88%, you will earn a B grade since both are equally a B? This is what the means comparisons analysis is like. We can see that both treatments "earn" an A in Figure 6-30, and since they have the same letter, they are both statistically the same.

Why do we care about Tukey means comparisons? How does this help us make better decisions with our data? Since we want to make decisions quickly, would it not be helpful to have some method where we could see rankings of treatments? Is looking at the letters not easier than looking and comparing treatments' means? What if you wanted to change products, maybe one that you have been using for several years will soon become discontinued or go off the market. How would you be sure a new replacement product would be as effective as one you have used previously? In fertilizers, could we not look at Tukey or Student's t means comparisons of our final height data and see if there were differences that would help us make better choices? Let us do just that with the final height data from the fertilizer experiment.

To get back to the Fit Model page, click on the small x in the far upper right-hand corner of the screen, and close out this current JMP analysis of the initial height data. We want to change our dependent variable from initial height to final height so click on initial height in the Y box to highlight it, and then click on the **Remove** button (Figure 6-31) to remove it. You can then return to the left-hand column with your cursor and select **final height** and then the **Y button** again to insert it as the new dependent variable. Since we still want to have the fertilizer treatments be our independent variable, leave fertilizer in the **Construct Model Effects** box. Click **Run Model** to run a new analysis with the final height data as the dependent variable.

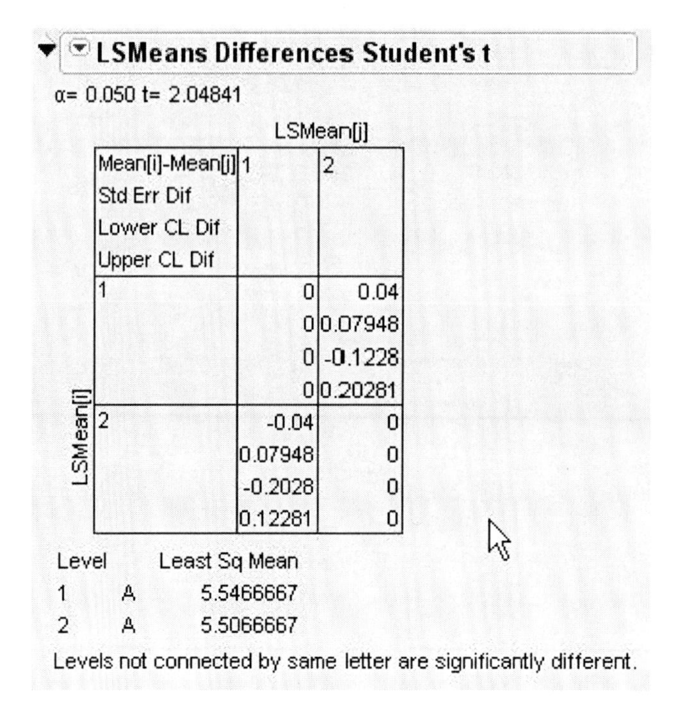

Figure 6-30. The LS Means Differences box from the Student's t-test shows there is no significant difference between the fertilizer treatment groups.

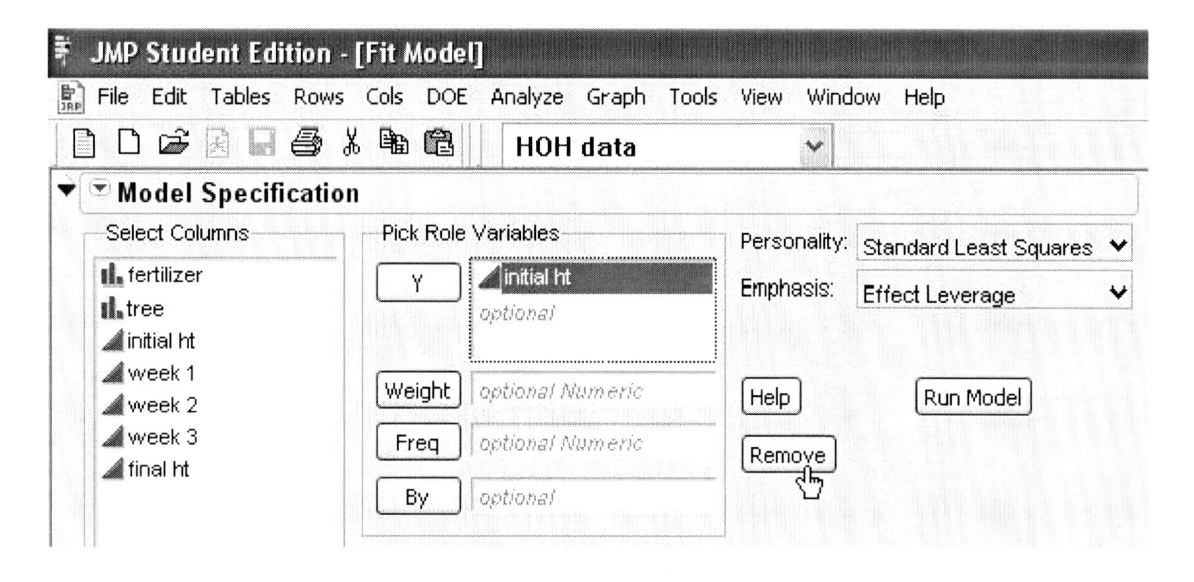

Figure 6-31. Changing the Role Variables from initial height to final height.

The **Actual by Predicted Plot** and **Leverage Plots** in Figure 6-32 have both been hidden from view since they will not contribute anything new to our discussion. You can hide unnecessary plots and graphs by clicking on the blue and grey triangles to their left. As seen in the ANOVA output from the final height data, the p-value appears on the left and you have the option of finding out more about the fertilizers in the right-hand column.

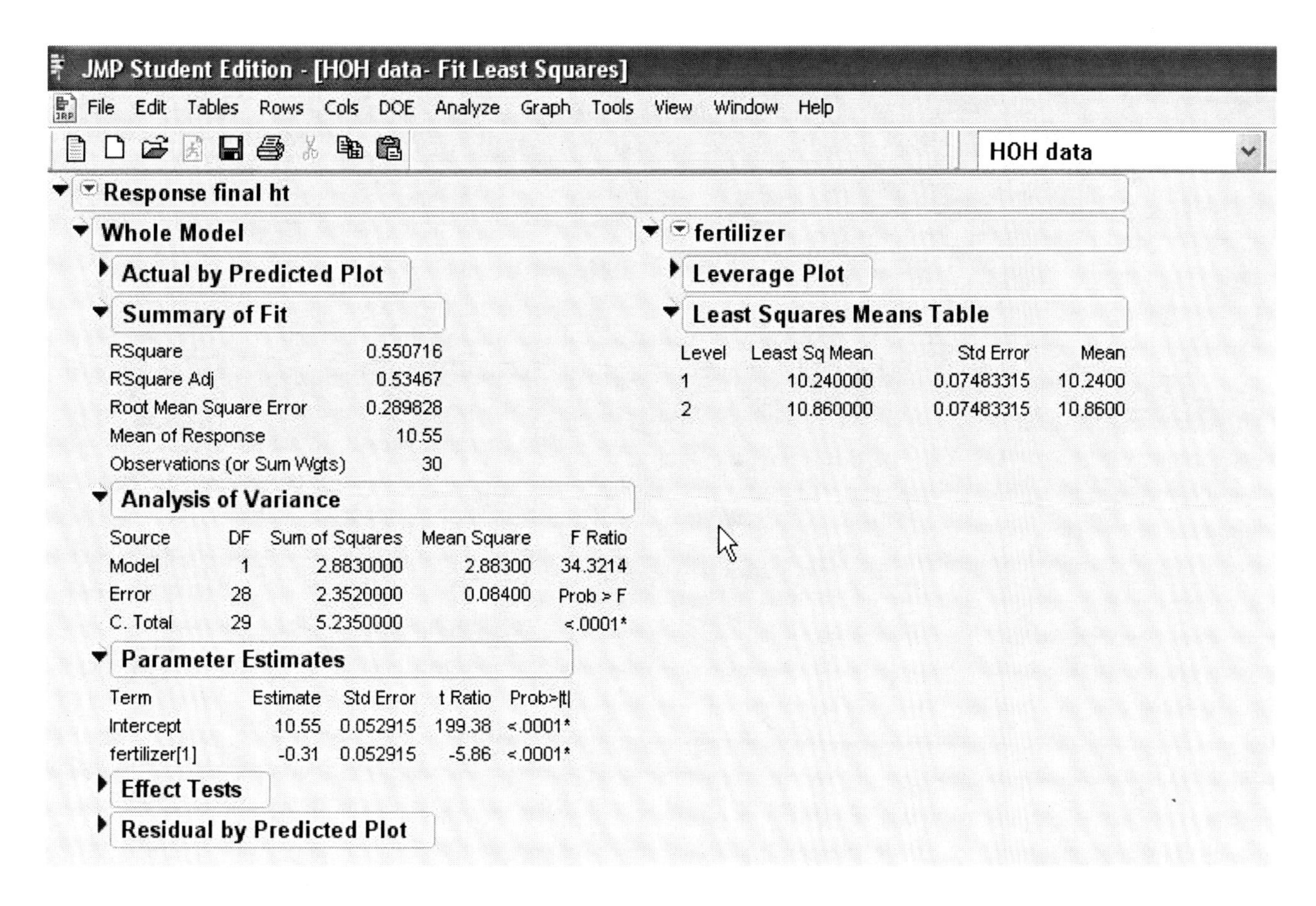

Figure 6-32. The Whole Model results give the final heights' p-value and ANOVA information calculated from the experiment.

We should be pleased to see a significant difference between our fertilizer treatments from the final height data (Figure 6-33). The p-value is 0.0001, well below our preset alpha of 0.05. Now that we have confirmed (from both Excel and JMP) that we have significant differences between our treatments, we might want to know which treatment had the greatest final height. This is easy with only two treatments, but if we had three or more it would not be so clear to see.

We do have the **LSMeans Table** in the right-hand column and it gives us the means of both treatments, 10.24 for fertilizer 1 and 10.86 for fertilizer 2. This confirms what we calculated for the means from each treatment with Excel back in Chapter 4.

Again, we can create a LS Means plot to show us a quick graph of the final heights plotted against each other. Figure 6-34 reminds us of the steps to creating the plot and Figure 6-35 shows the actual plot of both fertilizers. This plot shows a much greater difference between the two fertilizer treatments; can you see there is a sharp slant between the treatment means? Fertilizer 2 had a greater overall height gain than the plants in the fertilizer 1 treatment.

Now that we know that fertilizer 2 plants grew taller than those in the fertilizer 1 group, can we calculate the Student's t or Tukey means comparisons? We sure can; remember how to do that? Go back to the fertilizer heading and scroll down to Student's t (or Tukey, if you have three or more treatments) as in Figure 6-36. The means comparison results in Figure 6-37 show significant differences between the fertilizer treatments by placing fertilizer 2 in group A and fertilizer 1 in group B. Different letters represent significant differences and that makes it easy for us to look at the data and conclude those results. The treatment means are close, both in the tens, so it would be difficult to determine if they were different just by looking at the values. The letters make telling them apart so much easier (Figure 6-37).

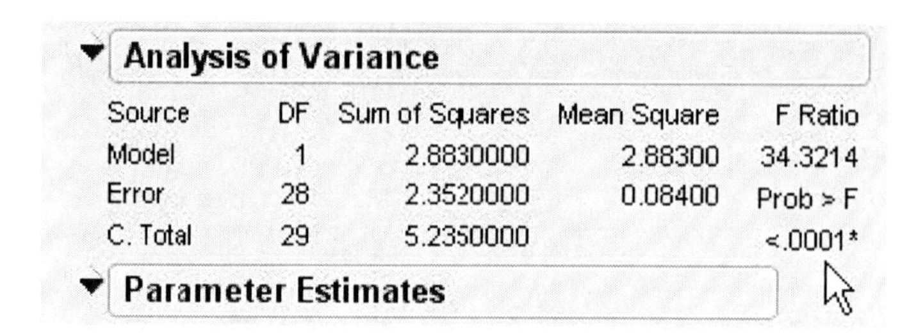

Analysis of Variance

Source	DF	Sum of Squares	Mean Square	F Ratio
Model	1	2.8830000	2.88300	34.3214
Error	28	2.3520000	0.08400	Prob > F
C. Total	29	5.2350000		<.0001*

Parameter Estimates

Figure 6-33. The ANOVA information calculated from the final height data show significant results with a p-value of 0.0001.

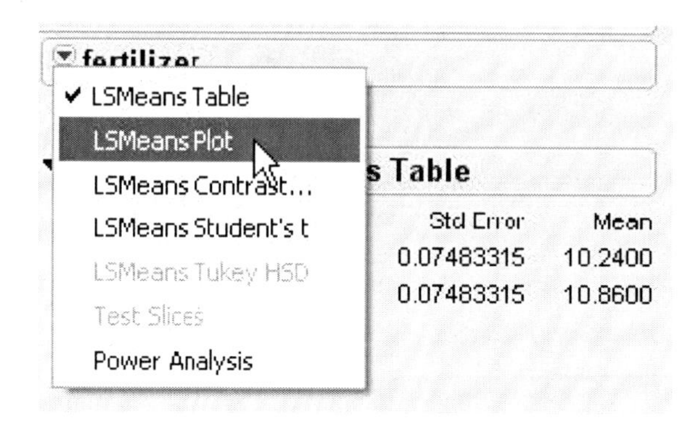

Figure 6-34. The LS Means Plot gives us a graphical view of our final data.

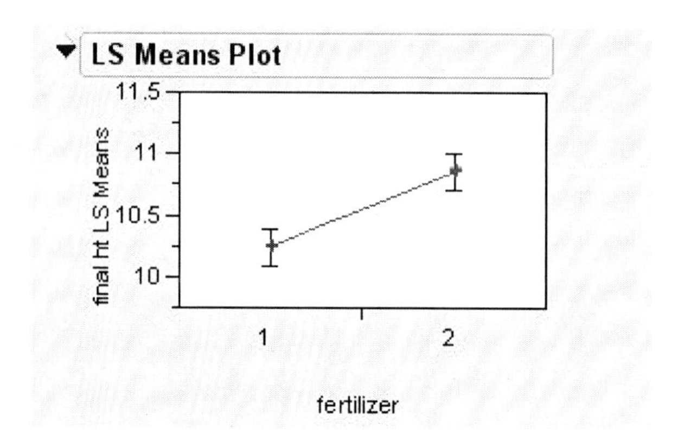

Figure 6-35. The LS Means Plot shows that fertilizer 2 made greater gains in height over fertilizer 1.

That is it for determining statistical significance (p-values) and learning about multiple means comparisons (Tukey). You now know how to compare data and determine if their treatments' means are statistically different or the same and can make better decisions about products, methods, or procedures with that knowledge. Our fertilizer data only open up more questions: should we change fertilizers from the number 1 brand to the number 2 brand since number 2 contributed to greater heights? Since fertilizer 2 worked well with foliage plants (weeping fig in our example), would it work as well with flowering plants, vegetables, or fruit trees? Is there a price difference between both brands and is using fertilizer 2 worth changing to?

These and other questions are very important and can lead to additional research projects. One thing we would like to point out to you before you get going onto another project is that we should look at the correlation between the fertilizer and the final heights.

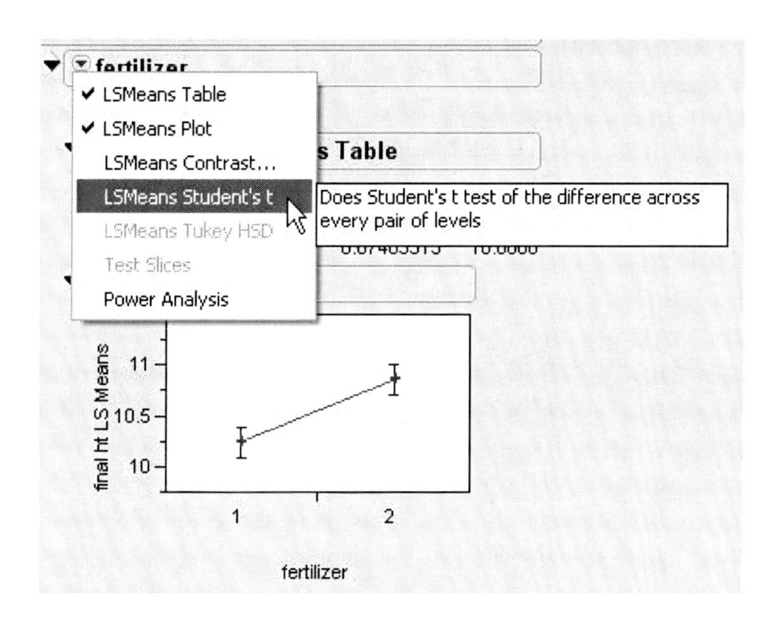

Figure 6-36. Calculating a means comparison for final height begins with selecting either the LS Means Plot or Student's t-test.

LSMeans Differences Student's t

α= 0.050 t= 2.04841

LSMean[j]

Mean[i]-Mean[j] Std Err Dif Lower CL Dif Upper CL Dif	1	2
1	0 0 0 0	-0.62 0.10583 -0.8368 -0.4032
2	0.62 0.10583 0.40322 0.83678	0 0 0 0

LSMean[i]

Level		Least Sq Mean
2	A	10.860000
1	B	10.240000

Levels not connected by same letter are significantly different.

Figure 6-37. The LS Means Student's t means comparison for final height confirms the significant difference between treatments by placing fertilizer level 2 above fertilizer 1.

Using Correlation Coefficients and Linear Regression to Correlate Variables and Make Future Predictions

As you were looking through the JMP output screen for the final height fertilizer data (6-32), I hope you noticed that in addition to the data being significantly different (p-value = 0.0001), there was the Summary of Fit section (Figure 6-38). In the Summary of Fit, there is an important item that may someday be relevant to your agriscience research.

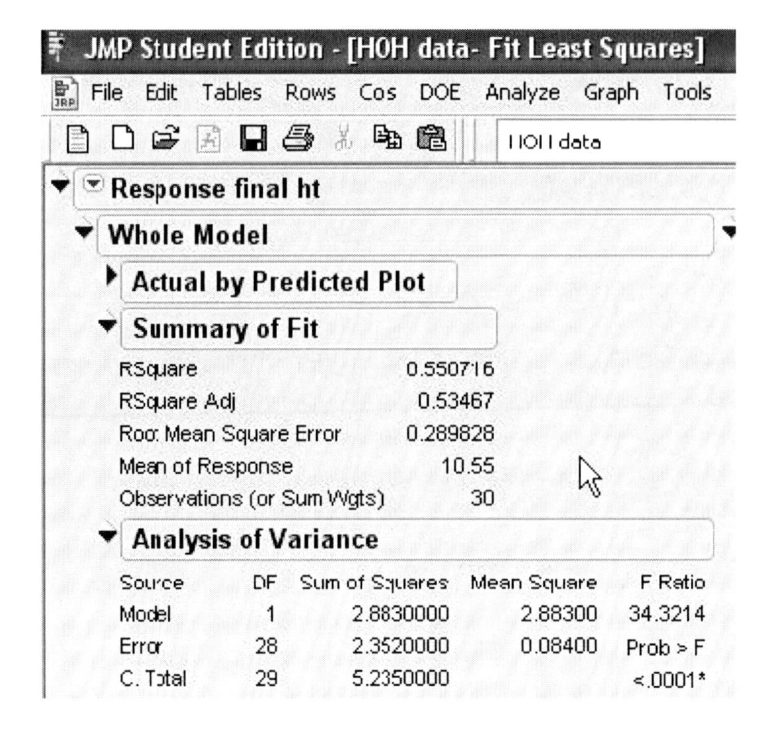

Figure 6-38. The Summary of Fit table provides correlation or association results from the data with the RSquare Adj value.

This new analysis item is the **RSquare Adj**, which is also known as R^2 adjusted. This is the percentage of variation in the dependent variable accounted for by the independent variable. In Figure 6-38, the RSquare Adj is 0.53467, which means that 53.5% of the variation in final weeping fig plant height is attributed to fertilizer brand. This is not a strong association; a strong association would be 0.70 or higher, up into the 0.85–0.99 range. Still, it is a good thing to review so you know if the model is good or not. In this case, it is not that great; fertilizer alone is not contributing to plant height.

If you are looking at the independent and dependent variables in your study and see that as one variable increases so does the other, this is a **positive correlation**. The data normally appear on a graph from the lower left stretching out to the upper right. Figure 6-39 shows a positive correlation between fertilizer rate and rose bloom yield from data in an Excel scatter graph. Can you verify that it is a positive correlation by seeing which direction the trendline goes? As the independent variable increases, so do the values of the dependent variable. This group of data is very close to the trendline, which is why the R^2 adjusted of 0.9865 is so high. This typically does not happen with real data unless you have a very good model; this is just shown to give you the idea of what almost-perfect data look like.

Also found in Figure 6-39 are the values of the linear regression line ($y = 0.0514x + 0.1924$), right above the R^2 adjusted of 0.9865. There appears to be a strong correlation here with an R^2 adjusted of 98.6%, meaning that 98.6% of rose bloom yield can be attributed to fertilizer rate, which isn't surprising.

Remember back in Chapter 4 when we created a trendline in Figure 4-24 using Excel? We introduced regression and mentioned that as you are inserting a trendline for your data, you can right mouse click on the trendline to bring up a **Format Trendline** menu. If you selected the **Options tab to Display equation on chart** and **Display R-squared value on chart**, the equation for the trendline and the R-squared value would appear near the trendline, as in Figure 6-39.

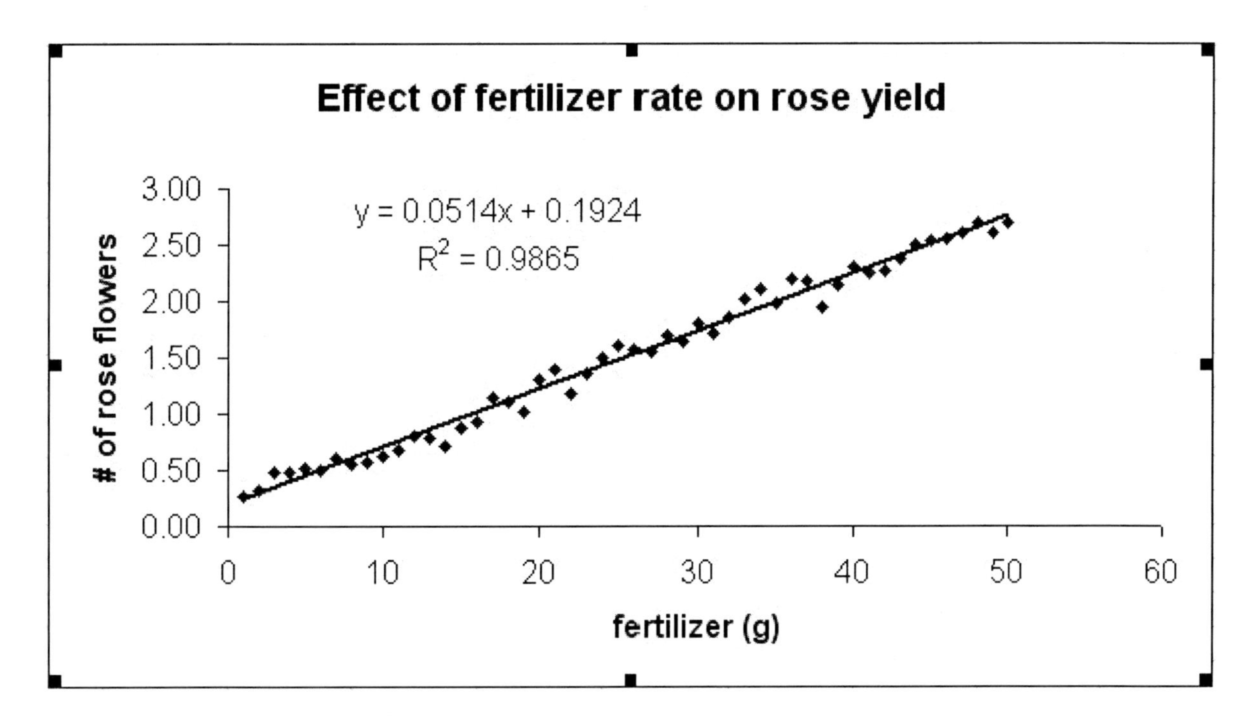

Figure 6-39. The trendline of the rose and fertilizer rate data has a linear regression line of y = 0.0514x + 0.1924 and an R^2 adjusted of 0.9865.

If you see that as you increase the independent variable and the dependent variable decreases, this is a **negative correlation.** A negative correlation is not a bad thing and there are several examples in agriculture. In crop production we see a negative correlation when we use plant growth regulators (PGRs) to shorten internodes on bedding plants and cotton. As we increase the rate of PGR, typically the internode length is reduced. Why would that be important in agriculture? Here are some other examples.

In animal science, we see a negative correlation when livestock are treated for external (and internal) parasites. Ranchers and livestock producers have to control flies, mites, ticks, internal worms, lice, and other parasites for the health of their animals. A producer does not want to hold back on this part of the management plan since the health and production of her/his herd (or flock or bands) depends on it! We also see negative correlations when we treat for weed, insect, or fungal infestation. When a fungicide is applied, the level of fungal pathogens is reduced.

Think about your own experience. When you give your pet flea medicine, the more you put on your pet (it is hoped up to the full dose you receive from your veterinarian's office) the fewer fleas your cat or dog will have on them. This is what we expect in a negative correlation, the greater the flea medicine dose (the independent variable), the fewer the number of fleas (dependent variable). Flea medicine dosage is usually calculated by the animal's body weight.

Since we are talking about R^2 adjusted, now is a good time to discuss regression and how it can help us predict future outcomes from model equations. Some people get correlation and regression confused. **Regression** is a statistical procedure to see if there is a connection between independent and dependent variables and suggests a straight line based on your data. After you have determined the regression line, you can make additional predictions using your actual data about future outcomes.

Figure 6-40 shows a negative correlation since the increasing doses of flea medicine applied to cats (in microliters, uL) reduces the number of fleas found on them. Suppose you wanted to know if using half the dose of a flea medicine on your cat would be a good idea to save you money yet still be effective. Could you use the linear regression line to predict the number of fleas you would find on your cat at various dosages?

You sure could if you have the line from the linear regression model (determined by Excel, it was **y = –0.0529x + 20.966**) and you were given the data range that it covered (it was 0 uL to 345 uL of flea medicine, taken from the graph). By the way, where do you think these data would come from and why? It could have been gathered by some local veterinarians in the area who care about pets and want to educate the public about using full dosages of flea medicines. Anyway, back to the flea problem.

Do you know what each number contributes in the regression line? The y is the number of cat fleas that would be on the cat by using a particular dose of flea medicine. The –0.0529 next to the x is how much y changes with each dose of x. What this means is*:

$$y = -0.0529x + 20.966$$
$$\text{\# of cat fleas} = -0.0529(\text{uL flea meds}) + 20.966$$

**For every additional unit (uL) of flea medicine you give a cat, there is a decrease in the number of fleas by 0.0529. It is a decrease because the slope (-0.0529) is negative.*

So, if you want to give your cat 200 uL of flea medicine, you could work out the number of fleas you would expect to find on your cat. Plug in 200 for x and solve:

$$y = -0.0529(200) + 20.966$$
$$y = 10.58 + 20.966$$
$$y = 31.546\text{, or in real numbers, about 32 fleas.}$$

Now, plug in 345 for x, the full dose of the flea medicine:

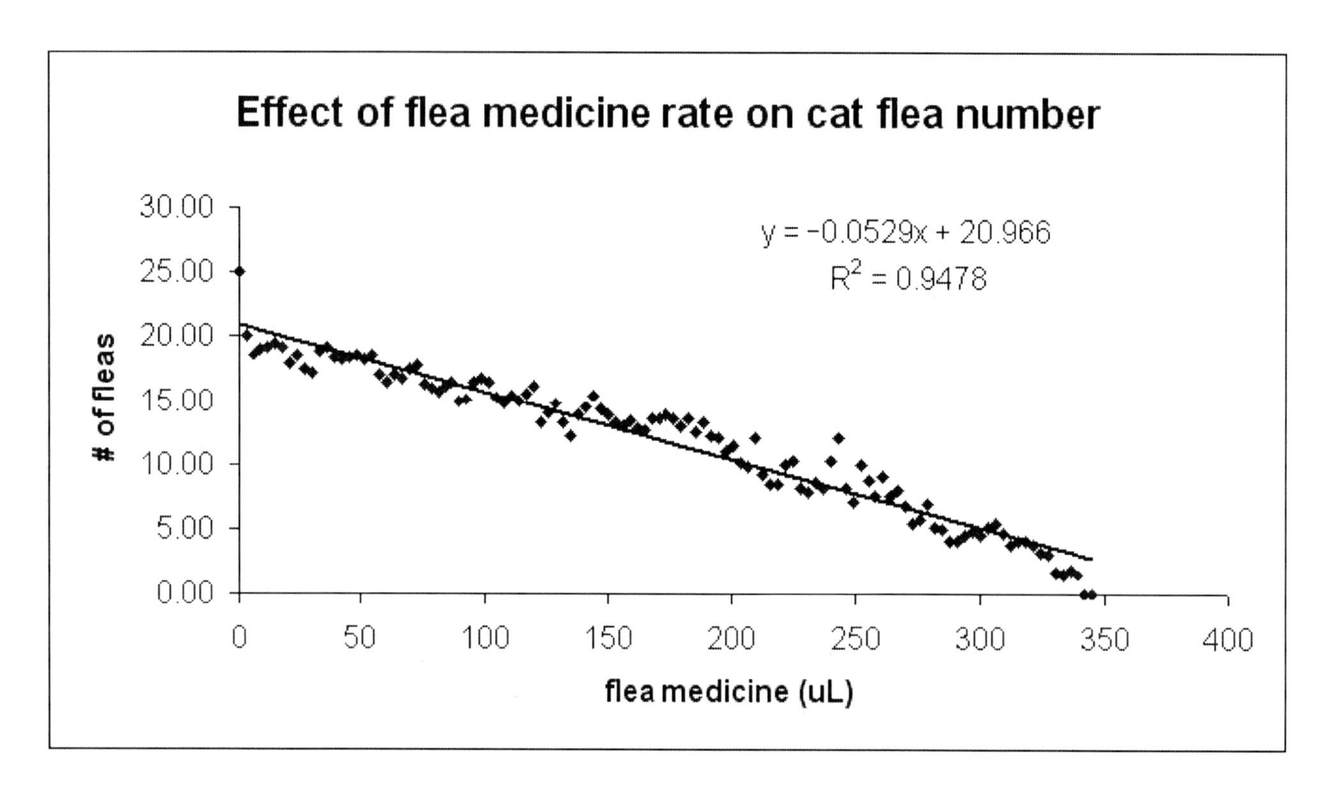

Figure 6-40. The linear regression line of the flea medicine data is y = –0.0529x + 20.966 with an R^2 adjusted of 0.9478.

$$y = -0.0529(345) + 20.966$$
$$y = -18.2505 + 20.966$$
$y = 2.7155$, or in real numbers, about 3 fleas.

Wow, 32 fleas compared to only 3 fleas. It makes sense to use the full dose of 345 uL of flea medicine on your cat instead of only 200 uL!

Correlation is a very helpful tool to see if there is a strong linear relationship between the two variables and can range between +1 to –1. A positive value shows a direct association between the variables and a negative value shows an inverse relationship. From Figure 6-40 we saw the correlation between flea medicine dosage and flea number, 94.8%! That is very good and tells us we need to pay attention to these data because they are strong and we could probably learn something from them.

Correlation values close to 0 indicate very weak or no association or correlation between the variables. When we saw our fertilizer RSquare Adj was 0.53467, that would be saying that there is some association between the variables but it is not very strong. Remember, all the correlation or association tells us is that the variables are associated with one another—it does not mean that one causes the other. We cannot say anything about cause and effect from the relationship but we can sure take a second look at the data to see if we can determine the cause. The flea medicine data are almost a freebie, the medicine is what is controlling the flea numbers, we can almost bet on it. However, for other data, we would have to look at all other variables and environmental factors.

Looking at the R^2 adjusted and p-value will help tell you more about how you are analyzing the data from, say, the fertilizer experiment on the weeping fig trees. Yes, there was a significant difference between the two fertilizer treatments but it was small (when you look at the treatment means). This is probably why the R^2 adjusted was weak.

When would we use the term associated or correlated? When you have numerical or quantitative data that have an RSquare Adj of 0.70 or higher, you would have a **correlation.** If you have qualitative data (from ranks or categories) that have an RSquare Adj of 0.70 or higher, that would be called an *association.* There are several examples where correlation or association would help you make decisions about your cultural practices:

1. Are artificial insemination techniques associated with pregnancy in dairy cattle?
2. Does herbicide strength correlate with weed control?
3. Do homework or lab scores correlate with exam scores?

You get the idea. If your correlation score ends up in the 0.90s, that is very good and says that you are doing something right, check the means comparisons (Tukey) on JMP to see which one gives the highest/best results overall.

As for regression, it is a very important analysis tool to make predictions based on previous experiences or data collected. However, it is very important that you realize that you can only use a regression line to predict dependent variables if the independent variable you want to use is included within the original data points from the regression line. You cannot predict y values from x values that were not included in the original range of data. This is like going "out of bounds" from the original data (the original boundaries) and is called **extrapolation.** Extrapolation is a big problem because you cannot be sure if your x value is "out of bounds" or the original data would have resulted in that same regression line. We see this all the time in advertising or in commercials. It is "stretching the truth" to make a product look better or to include it with a claim made by someone else.

Be aware when looking at data, you now have some more tools to use when seeing if advertising claims are justified or if they are extrapolating to sell a product.

Using X^2 (chi square) to Determine If Count Data are Significant

Sometimes our data are quantitative in nature but are the result of count data. Count data are those that come from just that: number counts or tallies! Whether it is number of lambs born, number of seeds that germinated, number of fruit trees that set fruit, or number of Jersey dairy cows that are pregnant, we can measure it by counting the results.

Some experiments involve the use of count data and gather information depending on presence or not (insect infested vs. clean, germinated vs. not germinated, flowering vs. not flowering, pregnant vs. open females in livestock) or preferences desired (plain vs. peanut, almonds vs. walnuts, orange vs. grape juice).

You might find this type of data collection useful for preliminary experiments. The analysis is called a chi square analysis, or X^2. There are two ways to calculate significance with count data and you can use JMP software or you can access a contingency table calculator or a X^2 calculator from the WWW. An example from the WWW is from College of Saint Benedict and Saint John's Universities: http://www.physics.csbsju.edu/stats/contingency_NROW_NCOLUMN_form.html. This is a good opportunity if you want to learn about website calculators and their availability online. You can find just about any online calculator to compute some value for you, including conversions from one type of measurement units to another (uL to mL, pounds to grams, etc.). Do not get too hung up on the term "calculator," it represents specialty software programs, (not like the little solar calculator or the advanced graphing calculator you have in your backpack), some of which are very sophisticated.

For both methods, either using the WWW or JMP, you will need the treatment types found in the experiment (e.g., light applied vs. no light application when comparing seed germination) and the number of seeds that germinated or did not germinate. Use of the X^2 website calculator will be discussed first.

For the website calculator mentioned at the web address example above, set up the chi square to have two treatments and two possibilities (the seeds either germinated or they did not) so from your table of data (Table 6-3), enter the number 2 into both row and column (there are two of each), and click **Submit**. You will see a contingency table that has two treatments (A and B) with the two possibilities (Figure 6-41). Have #1 be "did germinate" and #2 be "did not germinate," column letter A will be "light applied to the seeds" and column letter B will be "no light applied (the control treatment) to the seeds." Data for this germination experiment are found in Table 6-3 and are entered into the contingency table at the website (Figure 6-41).

Table 6-3. Count data recorded from a germination experiment testing the effects of light presence on seed germination.

Presence of light:	Light	No light
Number of seeds germinated	20	15
Number of seeds not germinated	25	13

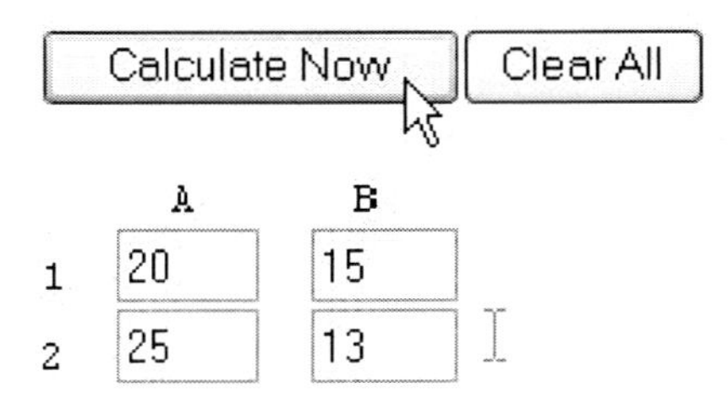

Figure 6-41. The data table from a X^2 website calculator.

chi-square = 0.576
degrees of freedom = 1
probability = 0.448

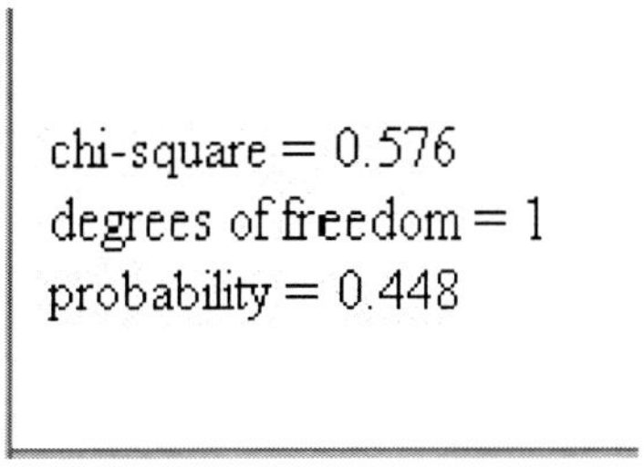

Figure 6-42. The data results from a X^2 website calculator shows the probability that there is no difference between the treatment groups. It is non-significant.

Click on the **Calculate Now** button (Figure 6-41), and you will see the contingency table with the actual and expected outcomes. At the bottom of the figure (Figure 6-42), look for the probability, in this example it is 0.448. Do not use the chi square value because it is a table value (which is used in computing the p-value). Since we want the probability (the p-value) that there is no difference in germination between treatments (i.e., no difference with presence or absence of light), use the probability value, p-value.

With such a high p-value of 0.448 (which is greater than 0.05), **our data are not significantly different and we fail to reject the null hypothesis** that there is no difference between treatment groups. Remember that this is like saying we accept the null hypothesis that there is no significant difference on germination no matter if we have light or no light on seeds during the germination process. You can conclude that having light on the seeds does not make a difference on increasing seed germination so it does not make sense to waste energy, so turn the light off.

Sometimes website calculators are not always available, the website goes down, the server is down, or something else happens disallowing its use. It may be possible that this website calculator may not be available to you in the future (the author retires, a new website is created, the link to the old website is removed, etc.). If this happens, you can use JMP to determine the X^2 p-value and/or can compare this p-value to the one found on a website calculator for insurance (and confidence!).

On a new data sheet in Excel or in JMP, enter the data as seen above with the treatments at the top and frequencies found at the end of the experiment and import it into JMP (Figure 6-43). Note the columns are changed to nominal for germination and treatment; this is done by clicking on the column titles and changing the data and modeling type. The number germinated column remains a numeric, continuous variable. You can use coding variables (as used in previous examples) or you can type in the words for the treatments as distinct categories. For small data entry such as this example, labels (e.g., did and did not) work out well and are easy to list. The treatments for this germination test were presence of light on the seeds or no light present.

To analyze this type of experiment, go to **Analyze, Fit Y by X** (Figure 6-44), and you will find the next distribution dialog box. Place your independent variable treatment into the **X, Factor** box (Figure 6-45),

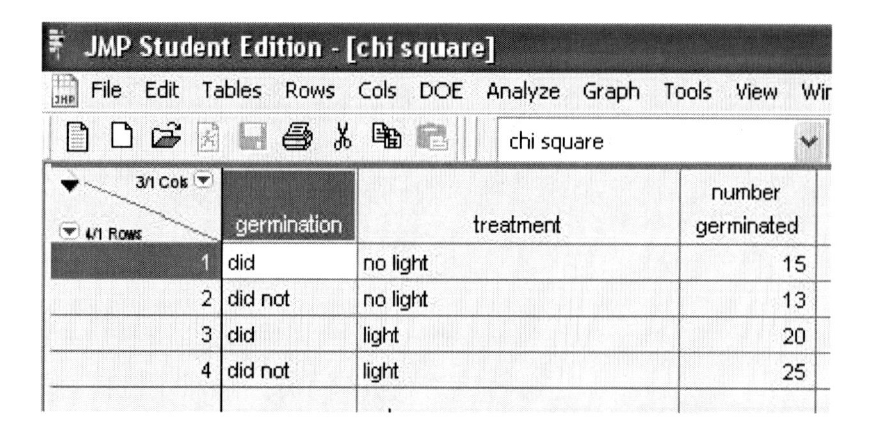

Figure 6-43. Germination treatments entered in JMP can be described with words or coding variables (1, 2, and so on). In this figure, they are listed as words (categories).

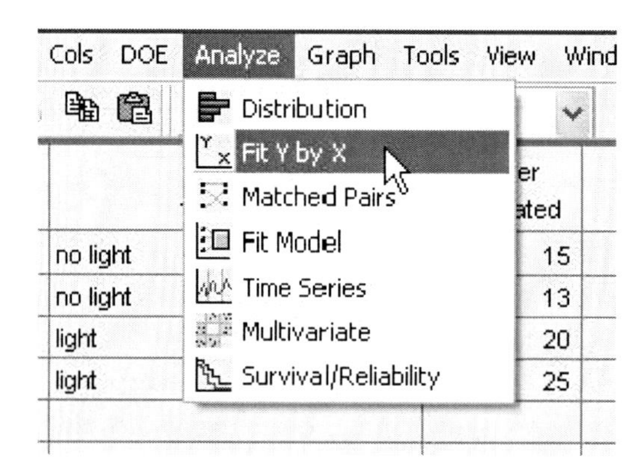

Figure 6-44. Starting with the Analyze menu in JMP provides a starting point on the selection of the procedure to use. Fit Y by X can be used to compare two means from two samples. Fit Model can be used to compare three or more means from many samples.

the dependent variable germination into the **Y, Response** box, and the number germinated in the **Freq** box. Click on **OK** to analyze the data. It is important that you do not forget to include the number germinated into the frequency box since it is the germination data (count data) that is the frequency we are analyzing.

The X^2 analysis of the data (Figure 6-46) shows the percentages of seeds obtained from each column and row and the p-value (from the Pearson test) for the data from this experiment. The Pearson p-value of 0.4478 obtained by the X^2 in JMP analysis is the same value we determined using the X^2 website calculator and is not significant, there are no differences in seed germination between the treatment group and the control group.

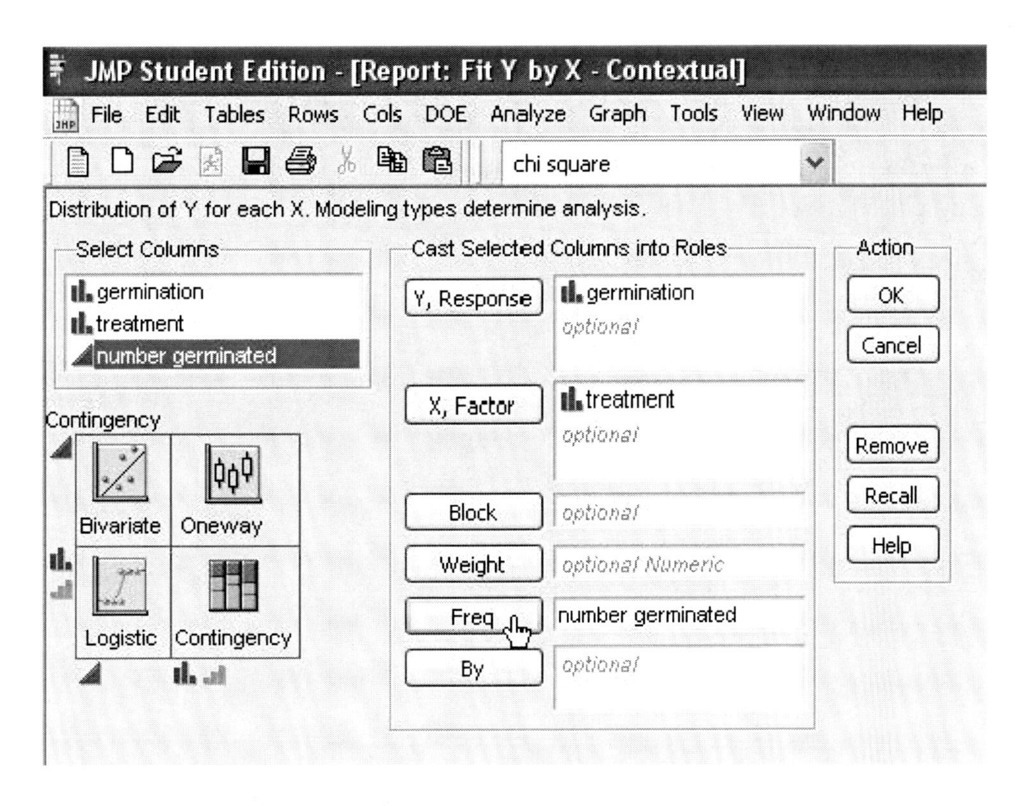

Figure 6-45. Entering the manipulated and responding variables in JMP is done by placing each of the variables in the Y, X, and freq boxes.

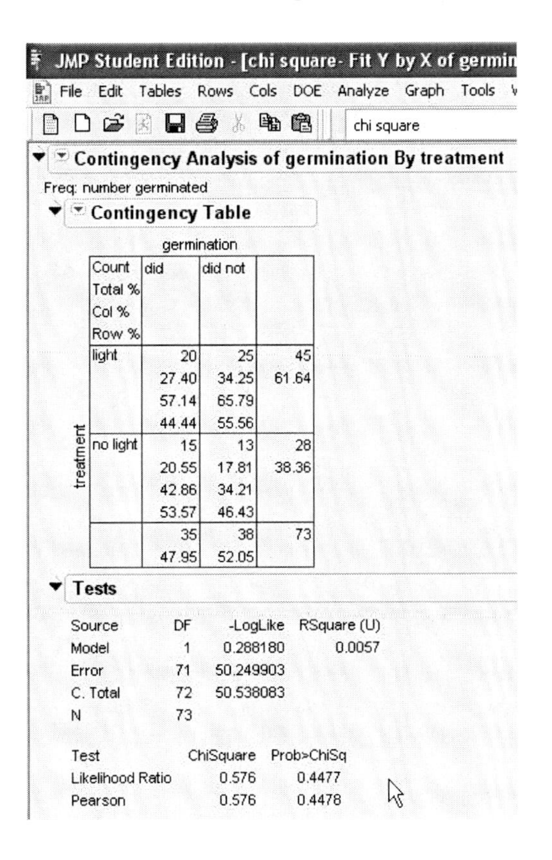

Figure 6-46. The Pearson p-value for a chi square test performed with JMP is found at the bottom of the analysis, and is 0.4478, a non-significant value.

It is helpful to note that there were different numbers of seeds planted in each treatment in this chi square example, and that is what makes this a useful test. You might not always have the same counts to compare, and that is alright. The treatment group had 45 seeds planted compared to the control group which only had 28 seeds. It is not always possible to have the exact number of replicates in each treatment group although we try to have them as close as possible. Having the contingency table can be a handy tool when you want to determine percentages of groups or treatments that resulted from the experiment.

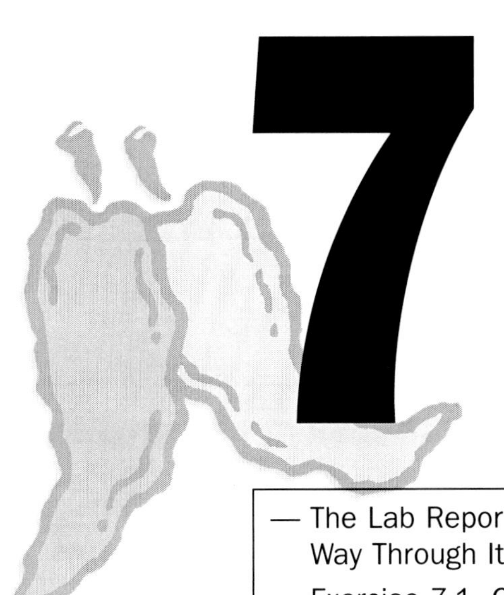

7 Developing the Final Lab Report or Scientific Paper

The Lab Report: Start with the Introduction and Objectives and Work Your Way Through It

Now that you have organized your references for your literature review, kept thorough notes from the development of your experiments, and reviewed the tables, figures, and images that you recorded for your results and discussion, it is time to get down to writing the lab report!

Remember to review Chapter 2 for hints and tips about gathering your materials and take time to review several of the journal articles posted on our class website. Learning how to compose scientific writing is just like learning how to ride a bike, see what others are doing, practice, get advice, and continue out on your own.

To get you started with the introduction and objectives, work through Exercise 7-1. This exercise will assist you in composing the introduction by giving you key questions you should consider when formulating your thoughts. Remember, your background information is necessary to have an effective "selling point" for your reading audience, to get them to buy into your research and see why your objectives are important.

Make sure you note the topic and due date for this and all additional writing assignments. The specific paper format (font, font size, margins, page numbering, etc.) can be found in the course syllabus.

Each additional exercise will help you break the lab report down into manageable chunks or modules so you do not become overwhelmed by writing the entire paper in one sitting (which is virtually impossible anyway). Make sure you have the data from the class experiment that is being used in the lab report. As always, follow the directions for each assignment, noting there are examples in the journal articles on our homepage.

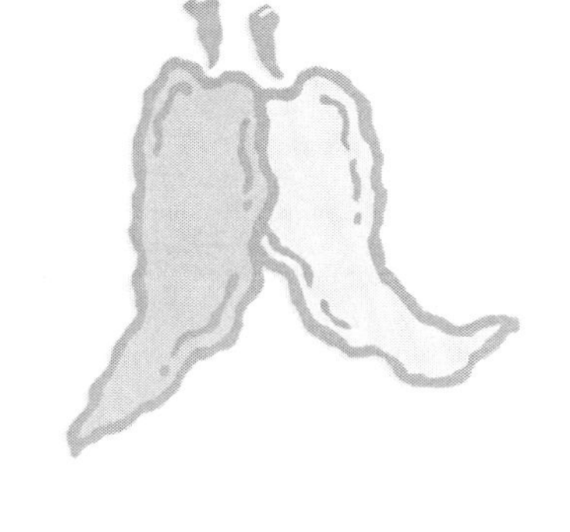

Exercise 7-1. Composing the Lab Report: Introduction and Objectives

Name Jake Philpott

Lab Section No. 05

Date 2-24-10

Being a successful student researcher means you have found a problem or phenomenon in the agricultural sciences and want to address it by designing a research study. To obtain grant funding or to sell your idea to your colleagues for support, you need to write an introductory paper that describes the nature of an agricultural problem, why it is important that this area be researched, what is known about the topic, and what you hope to accomplish by studying it.

The Assignment

For each of the following sections, follow the format found in any of the introduction sections from any of the posted journal articles in our class homepage. You may write the sections separately or together but remember the point of this assignment is to support your research and sell its importance of being studied to the reader.

Review Chapter 2 for suggestions about writing your introduction and objectives.

I. **Background for the Introduction**: What do we already know about this problem or phenomenon in agriculture? Cite other researchers (a minimum of 10 are required), how do they agree/disagree? Why is this problem important to study? Are there conflicting data that need to be clarified or are you trying out applying a current method or protocol to a new agricultural or horticultural crop? Are there similar studies out there in other disciplines or subfields?

II. **Objectives of the study**: Give 3 objectives you will address in this research. The objectives need to be quantitative in nature, something that could be numerically measured, so consider the independent and dependent variables you would include in this work.

III. **The topic for this lab report is** ___________________________. Your instructor will provide you with the overall agricultural topic for your lab report. It will encompass one of your research projects from the garden plots or greenhouse experiments.

IV. **The due date for this assignment is** ___________________________. This date should be easily located on the syllabus. Remember, no late work is accepted so get started on this assignment early.

Exercise 7-2. Composing the Lab Report: Materials and Methods

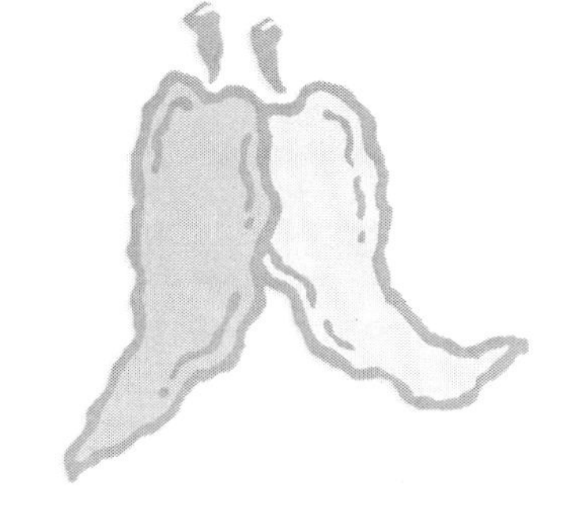

Name ____

Lab Section No. _____________

Date ______________________

Being a successful college researcher means you have learned how to create a materials and methods section in research reports and term papers. Using the materials and methods section from any of the posted journal articles on our class website as a guide, write a materials and methods section of your lab report that accurately and precisely describes the techniques and equipment/materials used in our agriscience experiment.

Remember this is not a list but is written in paragraph form as a written description of the procedures that were used to develop the experiment, measure the treatments, and terminate the plants. You need to compose this section in third person, past tense language.

This paper needs to be written in the format listed in the syllabus and will be double spaced.

I. **The topic for this lab report is** ______________________. Your instructor will provide you with the overall topic for your lab report. It will encompass one of your research projects from the garden plots or greenhouse experiments.

II. **The due date for this assignment is** ______________________. This should be easily located on the syllabus. Remember, no late work is accepted so get started on this assignment early.

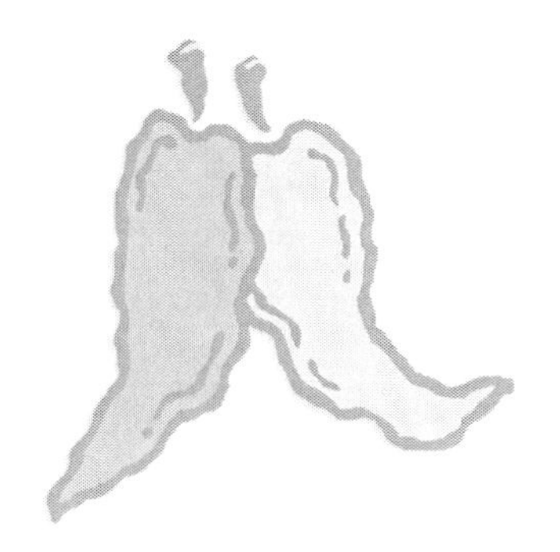

Exercise 7-3. Composing the Lab Report: Results and Discussion

Name Jake Philpott

Lab Section No. 05

Date ______________________

Being a successful student researcher means you have learned how to look at data, reported the relevant (significant) results to the reader, and discussed why you think the experiment resulted in the manner that it did. This part of the lab report should support or refute the null hypotheses (Ho) from your objectives. Use the numerical data you have collected from the lab experiments to create a graph and a table (both are required). Images (photos) of the treatments are optional but encouraged.

Review the steps of making a graph in Chapter 4 if you are unfamiliar with creating a graph in Excel. This paper needs to be written in the format listed in the syllabus and will be double spaced.

Hypotheses

Ho: There is no difference (or _____ has no affect on) ______________________________.

Ha: There is a difference (or _____ does affect) ______________________________.

The Assignment

For each of the following sections, follow the format found in any of results and discussion sections from any of the posted journal articles on our website. You may write the sections separately or together but remember the point of this part of the lab report is to support or refute the Ho with our data, pointing out those data to the reader.

I. **Results**: Report the data in a manner that is easy to understand. At least one graph and one table are required that you refer to in the text body of your paper. Don't forget to include the appropriate captions and references to your figures. Tell the reader what happened and what results you obtained from this research. Focus on your objectives that you wrote for your introduction section.

II. **Discussion**: Discuss your data as they relate to the hypotheses, i.e., do results from this study support our hypotheses or not? Were the methods adequate to test the hypotheses? Why do you think "we got what we got" from our research? Were there any limitations or problems that may have affected our results? Recommend a future research study that would be the next logical step of this research.

III. **The topic for this lab report is ______________________________**. Your instructor will provide you with the overall topic for your lab report. It will encompass one of your research projects from the garden plots or greenhouse experiments.

IV. **The due date for this assignment is ______________________________**.

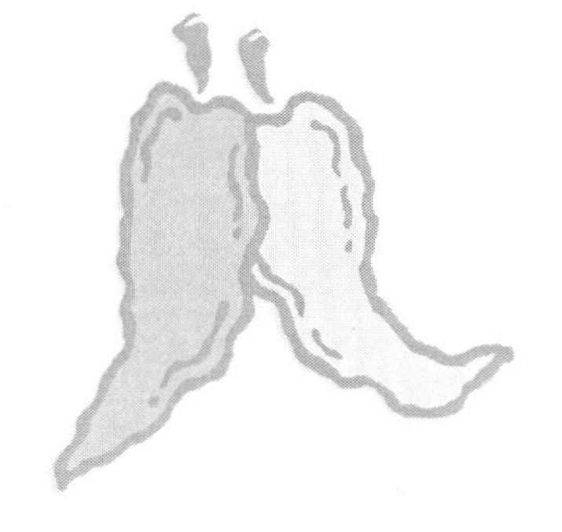

Exercise 7-4. Composing the Lab Report: Literature Citations

Name ______________________________

Lab Section No. ______________

Date ______________________

Being a successful college student/researcher means you have learned how to create a literature cited page in research reports and term papers so you can locate references and resources in the future. The following peer-reviewed journals (see list, below) are suggested for this part of the lab report. You task is to take the 10 references that you used to compose your introduction section of your lab report and put them into the correct literature citation format.

Word process your list of 10 references in alphabetical order by the primary author's last name and *follow the format from the Journal of American Society for Horticultural Science.* You will find citation examples for journals, books, and electronic formats from the literature citations included in posted journal articles and there is additional information on our class website.

This is only a list since you have already read the references and cited them in your introduction section. What is different between this assignment and what you see in journal articles is the following: *number your references 1–10, single space within each reference, and double space in between each reference. Follow the font size and type requirements as listed in the syllabus.*

The following journals are suggestions to get you started and there is no minimum number of citations that must come from each journal (e.g., you might have used only one of the following journals for all 10 references):

Agronomy Journal

Crop Science

HortScience

HortTechnology

Journal of Agronomy and Crop Science

Journal of Agronomy

Journal of Environmental Horticulture

Journal of Environmental Quality

Journal of the American Society for Horticultural Science

Plant Physiology

Soil Science Society of America Journal

The due date for this assignment is ______________________.

Exercise 7-5. Composing the Lab Report: Title and Abstract

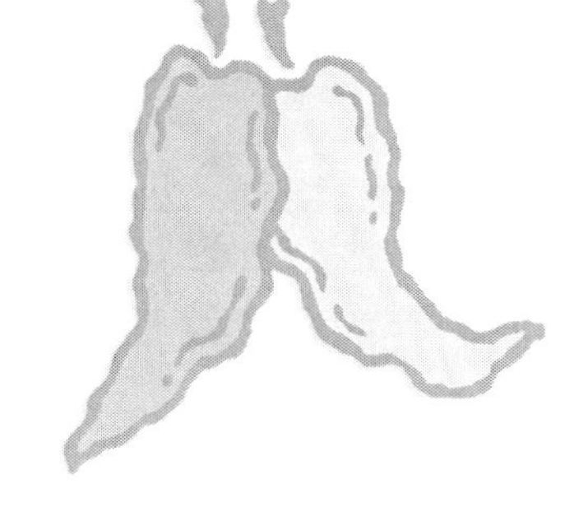

Name ____

Lab Section No. _____________

Date ______________________

Being a successful student researcher means you have learned how to create a title for a lab report or research paper, determine and list relevant index search words, and write an abstract of the work performed. This assignment will summarize the results determined in an experiment grown in the student garden plots or greenhouse.

For the title, describe the research performed based on your findings. The list of index words is useful for steering readers to your work in a search engine or database query on the WWW. Give 5 relevant index words that are pertinent to this research and list them below your title. Provide the abstract after the index words. The abstract should be concise and include details of the final differences between the plants found in the various treatments compared to the plants in the control group.

Word process your title, index words list, and abstract as per the formatting specifications in the syllabus. Follow the format found in any of the journal articles on our class website. Make sure to include the scientific names of all plants or other living organisms mentioned in the abstract, as well as any variety names and commercial names of products used. Place the title first, then the index words, then the abstract, double-spacing in between each section. *Double-space the abstract in a single paragraph.*

The due date for this assignment is ________________________.

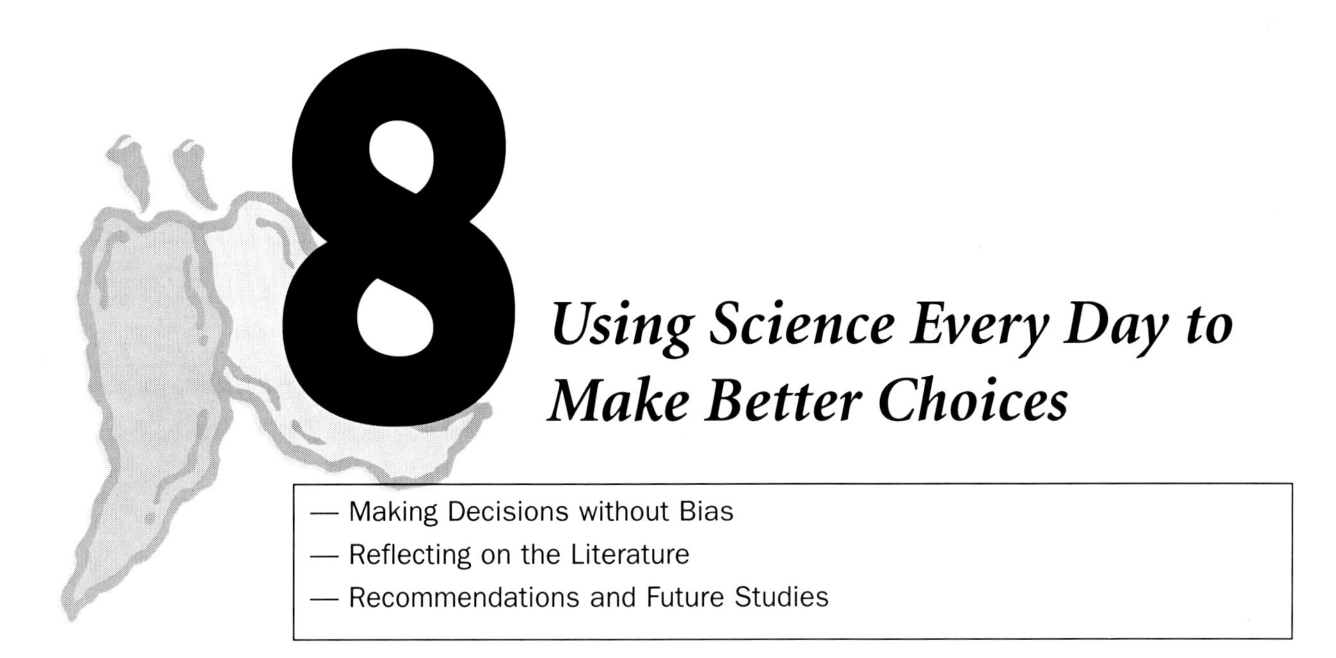

8 Using Science Every Day to Make Better Choices

Making Decisions without Bias

I hope it has become obvious to you how learning about science can help you make better choices every day just by looking at data. It does not matter what your background is, you have the capability to look at data, question how the data were collected, and determine if there is any bias involved with its reporting. Take advantage of what you have learned in this book to help you better present yourself (and your work) as you enter your career.

Reflecting on the Literature

Now that you know how scientific research is conducted, I hope you feel comfortable reading scientific journal articles and abstracts. There is a lot to gain from keeping in touch with the current literature, you can gain some insight on current trends and practices, and see what new products are out there to be utilized. You might even have an opportunity to use what you have learned in this class and connect it to your career objectives. By keeping up with the current literature in whatever discipline you happen to be in, you will be on the cutting edge of your field. Do not get lulled into the false security of "that's the way we've always done things," look at what the leaders in your field are doing. Perhaps you can continue with your science investigations to develop a new product, method, or procedure in your area of specialization on your own.

Recommendations and Future Studies

I hope that one of the most important things you have learned in this class is you should always learn from any activity, experiment, assignment, or experience. There is always something to be learned and, in the future, you just never know when you will need to have access to something or use something that you once did. A famous saying reads, "Learn all you can, that way you'll always know a little more than they think you know." This is very true, and is right on in today's ultra-competitive corporate world.

Remember that unbiased data can tell you something. Maybe it is just a confirmation of something you have already known or maybe it is something totally new and exciting; regardless, do not be afraid to act on your intuition. Think about how these results can make an impact on something else.

If it is positive, make a recommendation from it. Think to yourself, how can we apply this to another situation? Does this explain why this other event occurred? You just never know how the smallest suggestion or recommendation can turn into something really constructive or profitable, something really big.

I hope you enjoyed your semester with me in this course as much as I did. There are so many things to learn about agriscience, particular hortscience, and so many things I wish I could include in the course. But there is only so much time in one semester, so whenever you are around on campus, stop by and say hello and let me know how you are doing. I really enjoy hearing your success stories, how you have used agriscience in your life to better yourself and make decisions. Best of luck to you in your future!

9 Appendix

- — Designing a Poster Project
- — Other Experimental Designs and Arrangements in Agriscience Research
- — CRD Revisited with Hogs and Average Daily Gain (ADG)
- — RCBD Revisited with Nursery Blocks of Turf
- — LS Design Revisited with Dairy Cattle and Milk Yield
- — Interaction between 2-Way Factorials

Designing a Poster Project

After you have mastered your lab report research paper, you might want some other means to display your data and results for others to see. Figure 9-1 shows a common poster format that was created with PowerPoint® software from Microsoft Office®. This poster is 36"x 48" with a portrait orientation. What you cannot tell due to its grayscale color is how fresh and eye-appealing this poster is (when compared to a black and white term paper).

It has color images, color bars and heading backgrounds, color columns on the graph, and it has a personal touch by having the college's logo in the upper right- hand corner and color clipart of the university's mascot (in the upper left-hand corner). With the exception of clipart and colored textboxes, you already have the parts and pieces from your lab report to make a poster like this one. You have already made tables and graphs, you probably have images from final harvest, and you have the text written for each of the necessary sections. All you have to do is put them into a poster that you can make from a single PowerPoint slide.

To make a poster, start with a blank PowerPoint slide and set the size through the **File, Page Setup** menu. Typically the sizes range from 36"x 48" in either direction (landscape or portrait) or sizes can go up to 54". Be aware that having a bigger piece of paper does not make a better poster. Take care that you leave the background white. It is difficult to print a poster with a color background (and expensive too!) but more importantly, it is very hard to read.

Your headings should be legible from 6–7' away and have a take-home message clearly obvious from your title and figures. In term papers and lab reports, we like to see 12 point font size because it is easy to read. When you are making a poster, 12 pt. is too small; you will need to adjust your sizes for the readers who will be walking by glancing at it slowly. You want the font size large enough to make them stop in their tracks to read what you have presented. Here are some suggestions for poster font sizes:

1. Main titles should be 72–74 pt.
2. Text should be 28–32 pt.
3. Headings should be 40–48 pt. and can be bold

Play around with the sizes and color background on headings and subheadings to see what looks good to you. Do you want your name printed with first name first or last name first? Colored heading backgrounds can follow the color scheme of your school or university program. Insert the text boxes, title boxes, lines (from the **Drawing toolbar**), and figure boxes. Figures (pictures, graphs, and tables) should be converted to jpeg (.jpg) format prior to importing into the poster, (because they tend to get messy by doing strange things otherwise) and do not forget to include the captions! Use Figure 9-1 as a guide and you can also find hundreds of other posters on the WWW for inspiration and ideas. Figure 9-1 has lots of white space and is very uncluttered. The images and figures help break up text and have eye-appeal, especially when viewed in full color.

Poster presentations are a great way to convey your information in a quick, easy-to-read format that invites interested audiences to come in closer for more information and allows disinterested parties

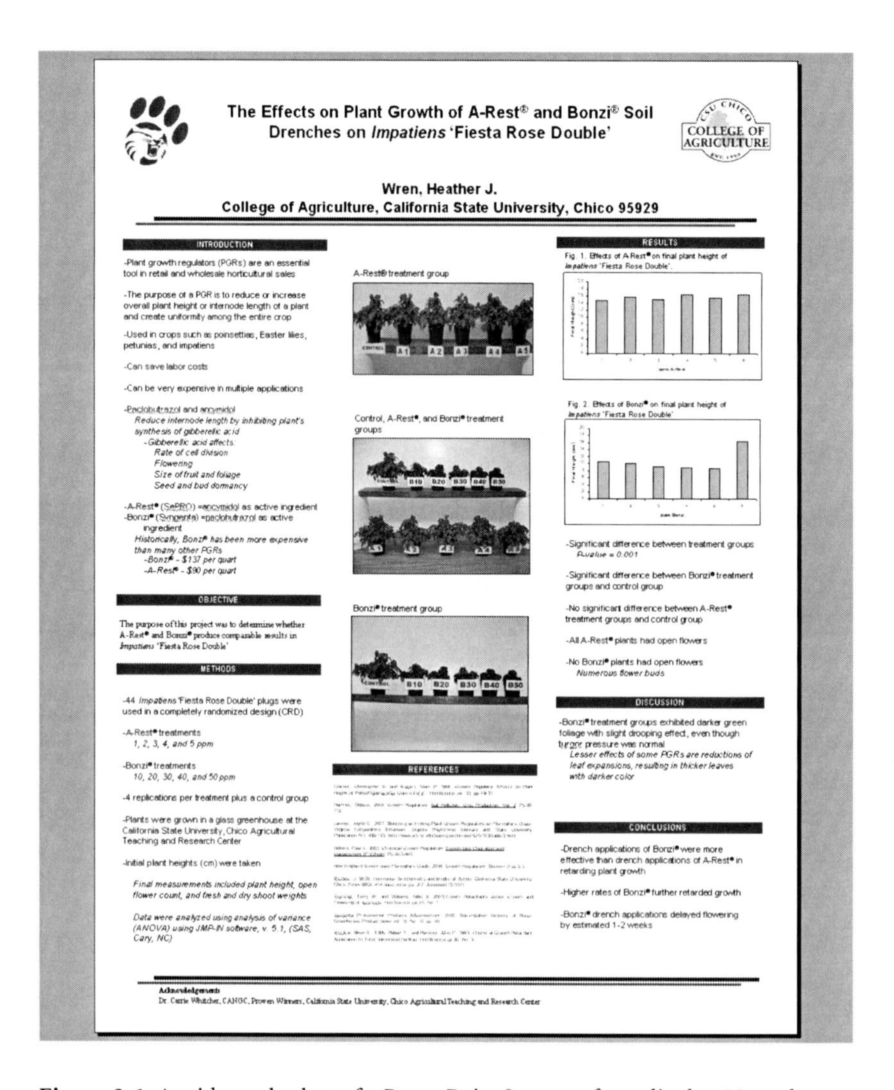

Figure 9-1. A wide angle shot of a PowerPoint® poster for a display. Note the use of white space and double spacing between paragraphs to avoid a cluttered look.

to keep their distance. Your limitation is your imagination; have fun and be creative, yet be somewhat conservative depending on your audience.

Other Experimental Designs and Arrangements in Agriscience Research

For students who would like to conduct additional agriscience research projects with designs other than CRD, using Excel and JMP® SE (Student Edition) are good choices because they allow you to create your own data sheets which can be easily modified to include additional dependent variables and expanded to include additional replicate numbers. You also have your data in a format that is easily accessible to send to colleagues or research advisors since most personal computers have Microsoft Office.

Back in Chapter 4 you learned how to create data sheets in Excel and import them into JMP for analysis (or analyzed the data in Excel directly). The goal of this section of this chapter is to show you additional data sheets for the experimental designs discussed previously, what the results are from some sample data, and explain the results.

CRD Revisited with Hogs and Average Daily Gain (ADG)

Since the completely randomized design is so popular in agriscience research, another example of a CRD comes from livestock and feeding trials. Determining which feed increases ADG is a profitable exercise for livestock producers because any dollar amount they can save on feed costs increases their bottom line. Figure 9-2 shows some ADG data from hogs fed four different diets with four replicates/treatment. We do not know if the control was the standard feed they had been eating previously but see we have data for ADG from four feeds. Quickly, can you tell what the dependent and independent variables are just by looking at the data?

Figure 9-3 shows the CRD hog feed data imported into JMP, note the symbols by diet and hog, they are both nominal types of data (named or coded) and ADG is labeled differently, it is continuous

Microsoft Excel - HOH Designs Arrangen

HOH CRD Example Data Entry

Diet	Hog	ADG
1	1	2.2
1	2	2.3
1	3	2.4
1	4	2.5
2	1	2.1
2	2	2.2
2	3	2
2	4	2.4
3	1	2.9
3	2	2.7
3	3	2.5
3	4	2.4
4	1	3
4	2	3.1
4	3	3
4	4	3.2

Figure 9-2. Data from an Excel file showing a CRD with hog feed and ADG.

(numerical). The analysis in JMP with ANOVA is started by selecting **Analyze, Fit Model**, which leads to the **Model Specification** box (Figure 9-4). Since ADG is the dependent variable, it goes in the Y box and both hog and diet go into the **Construct Model Effects** box. Do not forget to click **Run Model.**

What we can see from the data from Figure 9-5 is a significant difference in the overall model (p-value of 0.0012, by the white arrow). The effect tests show a difference between the diets (p-value of 0.0002) but no significant difference between the hogs as replicates (p-value of 0.6732). Even though we are not supposed to look at Tukey means comparisons if there is no significant difference, Figure 9-5

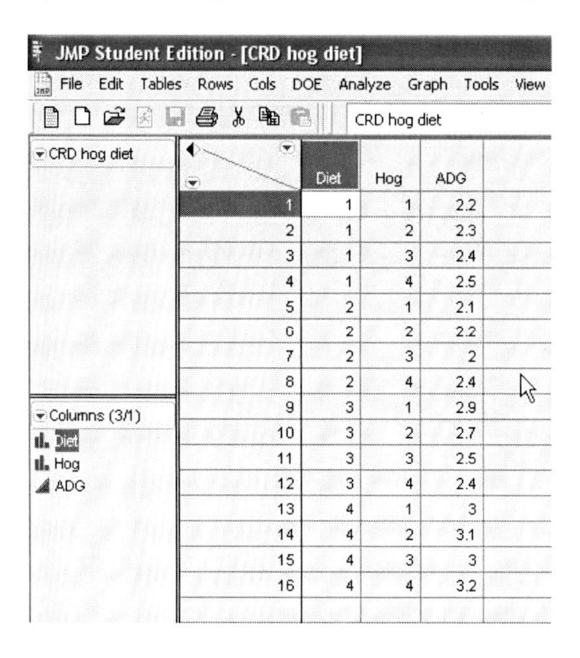

	Diet	Hog	ADG
1	1	1	2.2
2	1	2	2.3
3	1	3	2.4
4	1	4	2.5
5	2	1	2.1
6	2	2	2.2
7	2	3	2
8	2	4	2.4
9	3	1	2.9
10	3	2	2.7
11	3	3	2.5
12	3	4	2.4
13	4	1	3
14	4	2	3.1
15	4	3	3
16	4	4	3.2

Figure 9-3. Data imported into JMP showing a CRD with four hog diets, hog replicate number, and ADG from each replicate.

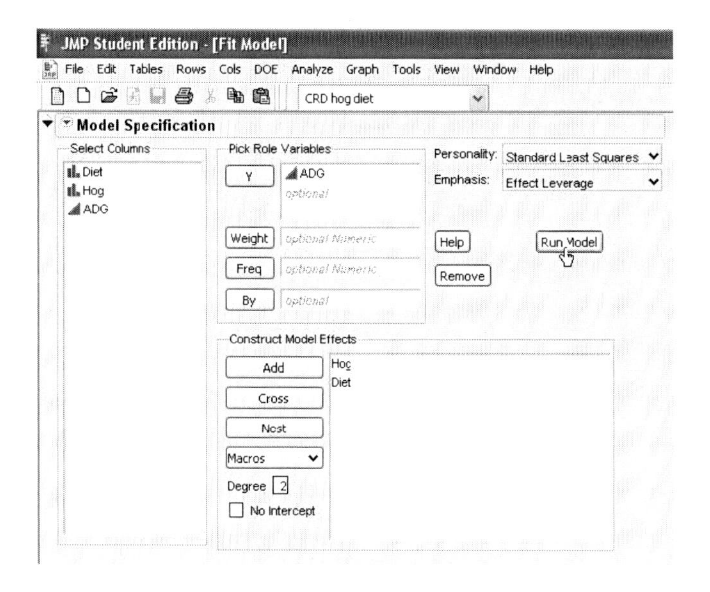

Figure 9-4. To run an ANOVA, place the dependent variable in the Y Role Variable box and the independent variables into the Construct Model Effects box. Click Run Model to analyze the data.

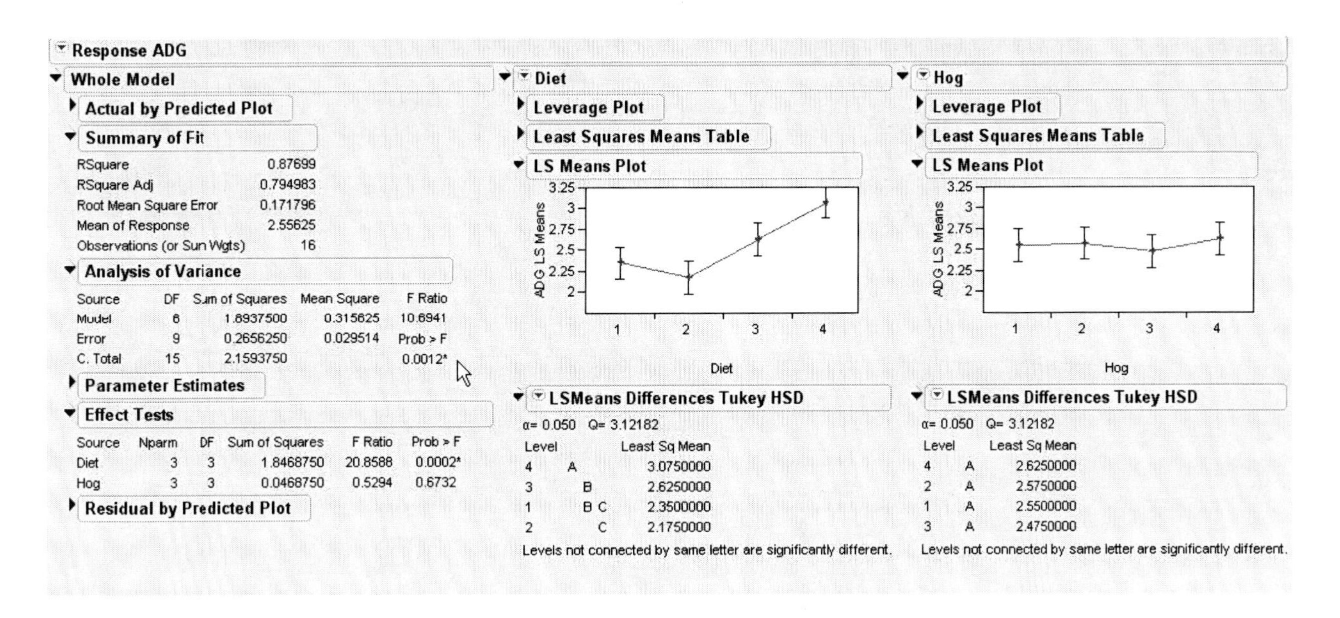

Figure 9-5. The final data analysis of a CRD with hog feed and ADG using JMP software.

confirms this lack of significance with capital letter As assigned to each of the hogs. Remember now, there were 4 replicates/treatment, which totals 16 hogs in this study (not 4 individual hogs consuming each type of feed). That type of study with rotations between only a few hogs would be a Latin square design, which we will see later on.

The Tukey analysis for the diets showed the differences with letters. Diet 4 resulted in the highest ADG and was significantly different than all other diets (it is an A, all by itself). Diets 3 and 1 were statistically the same (letter B) so they could be interchanged, and diets 1 and 2 were statistically the same (letter C), they too could be interchanged.

Figure 9-5 also gives the R^2 adj (0.794983), showing an association between the ADG and the feed. It is not the strongest association, but it is good. The take home message from this analysis could be any of the following:

1. If there is no difference in cost of the four hog diets, go with diet #4, it resulted in the highest ADG.
2. If diet #4 is too expensive, go with diet #3. If diet #3 is more expensive than diet #1, go with #1 since #3 and #1 are statistically the same as far as ADG.
3. Diet #2 should be avoided because it was not as effective as the other diets, resulting in 0.5–1 pound less ADG than the other feeds.

RCBD Revisited with Nursery Blocks of Turf

An example of a randomized complete block design comes from some nursery turf growing trials. Determining which fertilizer rate increases dry shoot yield is profitable for nursery growers because, like all growers or ranchers, any dollar amount they can save on inputs such as fertilizer increases their profitability. Figure 9-6 shows some dry shoot yield data from blocks of turf grass given five different rates of nitrogen fertilizer. There were five blocks studied in the turf field at the nursery so the grower wanted to know if location in the field affected the performance of the turf, in addition to the rates of N.

Figure 9-6 shows two formats of entering fertilizer rate labels, the one on the left gives the fertilizer rates as the actual rate in ppm N and the one on the right gives the fertilizer rate in coding variables. Either one works in JMP, it is up to you which format you prefer. Since we have already seen coding

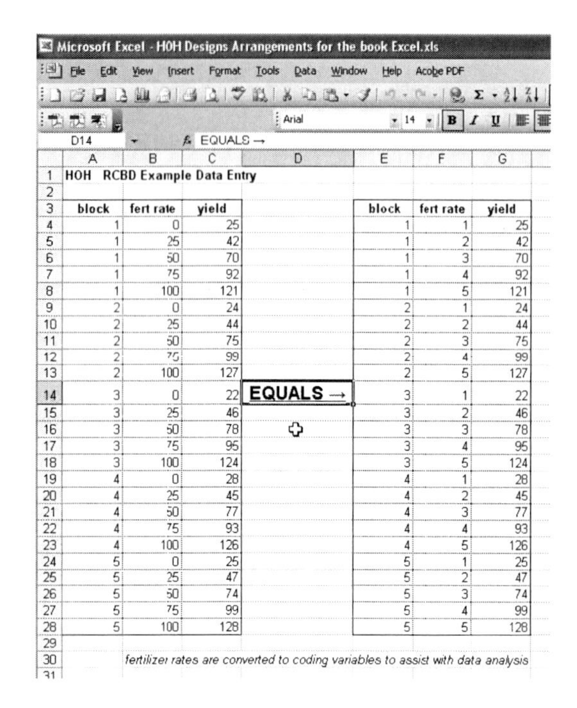

Microsoft Excel - HOH Designs Arrangements for the book Excel.xls

	A	B	C	D	E	F	G
1	HOH RCBD Example Data Entry						
2							
3	block	fert rate	yield		block	fert rate	yield
4	1	0	25		1	1	25
5	1	25	42		1	2	42
6	1	50	70		1	3	70
7	1	75	92		1	4	92
8	1	100	121		1	5	121
9	2	0	24		2	1	24
10	2	25	44		2	2	44
11	2	50	75		2	3	75
12	2	75	99		2	4	99
13	2	100	127		2	5	127
14	3	0	22	EQUALS →	3	1	22
15	3	25	46		3	2	46
16	3	50	78		3	3	78
17	3	75	95		3	4	95
18	3	100	124		3	5	124
19	4	0	28		4	1	28
20	4	25	45		4	2	45
21	4	50	77		4	3	77
22	4	75	93		4	4	93
23	4	100	126		4	5	126
24	5	0	25		5	1	25
25	5	25	47		5	2	47
26	5	50	74		5	3	74
27	5	75	99		5	4	99
28	5	100	128		5	5	128
29							
30		fertilizer rates are converted to coding variables to assist with data analysis					

Figure 9-6. Data from an Excel file showing a RCBD with turf grass and fertilizer rates.

JMP Student Edition - [RCBD]

	block	fert rate	yield
1	1	0	25
2	1	25	42
3	1	50	70
4	1	75	92
5	1	100	121
6	2	0	24
7	2	25	44
8	2	50	75
9	2	75	99
10	2	100	127
11	3	0	22
12	3	25	46
13	3	50	78
14	3	75	95
15	3	100	124
16	4	0	28
17	4	25	45
18	4	50	77
19	4	75	93
20	4	100	126
21	5	0	25
22	5	25	47
23	5	50	74
24	5	75	99
25	5	100	128

Figure 9-7. Data from a JMP file showing a RCBD with blocks in the field, fertilizer rates, and yield of turf weight.

variables in previous examples, Figure 9-7 shows the converted table in JMP with the actual rates in ppm N.

Figure 9-7 shows the RCBD turf yield data imported into JMP. Again note the symbols by block, fert rate, and yield. The first two are both nominal types of data and yield is continuous. The analysis for ANOVA is started by selecting **Analyze, Fit Model** (Figure 9-8) which leads to the **Model Specification** box (Figure 9-9). Since yield is the dependent variable, it goes in the Y box and both block and fertilizer rate go into the **Construct Model Effects** box. Do not forget to click **Run Model.**

What we can see from the data from Figure 9-10 is a significant difference in the overall ANOVA model (p-value of 0.0001, by the white arrow), an R^2 Adj of 0.996137 (99.6% correlation), and significant differences between fertilizer rates (p-value of 0.0001) and blocks (p-value of 0.0447). Note the blocks p-value is very close to our preset alpha of 0.05, which is nearing non-significance. Will this make a difference?

Let us start with the fertilizer rate data. Clearly there is a significant difference between treatments and by looking at the LS Means Plot in Figure 9-10, we see the highest fertilizer rate of 100 ppm N gave the greatest yield. The Tukey means comparisons confirms this significant difference by separating the treatments, with each treatment earning a separate letter (means difference). Each treatment was statistically different than the others, with the control group (0 ppm N) at the bottom of the treatments.

It is difficult to see any differences in the LS Means Plot for the five different nursery field blocks but the Tukey analysis for the blocks shows the means comparisons differences with letters. Blocks #2–#5 are statistically the same, they all have the letter A, and blocks #1–#4 are also statistically the same, they all have the letter B. This gives us an idea why the p-value for blocks in the effects test was nearing the α-level of 0.05. Block 5 was just a little bit different since it produced the highest yield and block 1 was just a little bit different since it produced the lowest yield.

Would this turf producer still use all five blocks interchangeably when growing new sections of turf? Probably so; in this case the differences between blocks was small, he or she would certainly want to

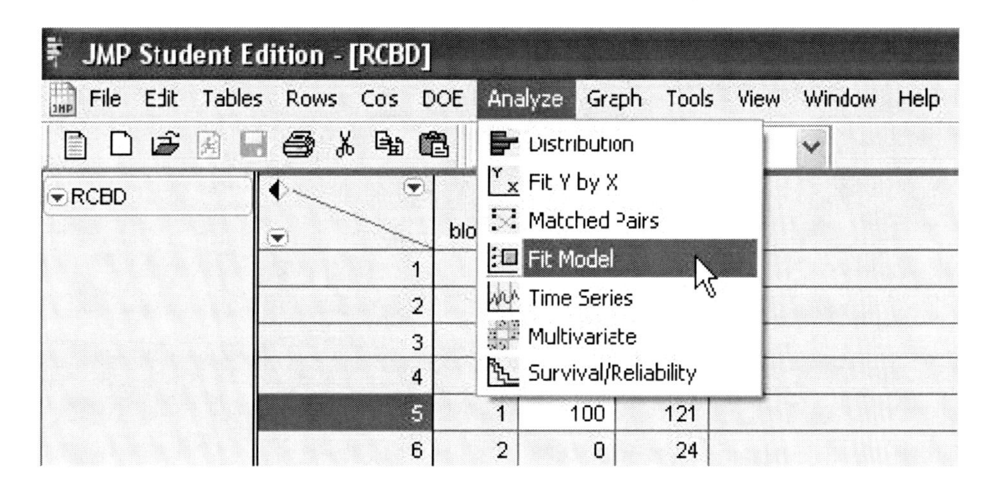

Figure 9-8. Begin the analysis of the RCBD by selecting Analyze, Fit Model.

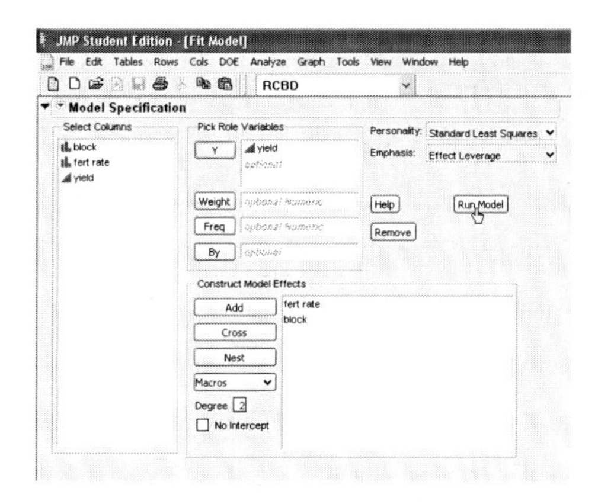

Figure 9-9. Place the dependent variable yield in the Y Role Variable box and the independent variables block and fert rate into the Construct Model Effects box.

determine why. If it is something permanent, like soil type, she or he might just add more fertilizer to plants in that block to compensate. The lowest block could have a greater amount of sand in the soil, causing greater levels of leaching and loss of nutrients, or there might be drainage problems in that section of the field. Testing the blocks will alert a grower to problems that might be easily controlled if tackled early on, before they get out of hand. Remember that in a RCBD the block is the homogeneous group, whether it be a turf plot growing on one soil type, a group of Hereford replacement heifers, or the garden plot amended with one type of compost. Each block (or homogenous group) of experimental units gets every treatment applied in that group, that way all treatments are analyzed.

The take home message from this RCBD analysis with turf grass plots could include the following:

1. There is a significant difference between the fertilizer rates that shows real differences in the field (look at the dry shoot yield). If quality and speed of production are important, which in commercial production they always are, go with the highest treatment. In turf production, sometimes money is no object. Think about the individuals who use turf fields (like golfers or athletes).
2. The grower might consider testing higher rates of N to confirm that this is the optimum rate for the cost. If the 100 ppm N rate performs the same statistically as say, 200 ppm N (as per Tukey), it would be an advantage to save money and go with the cheaper rate, the 100 ppm N.

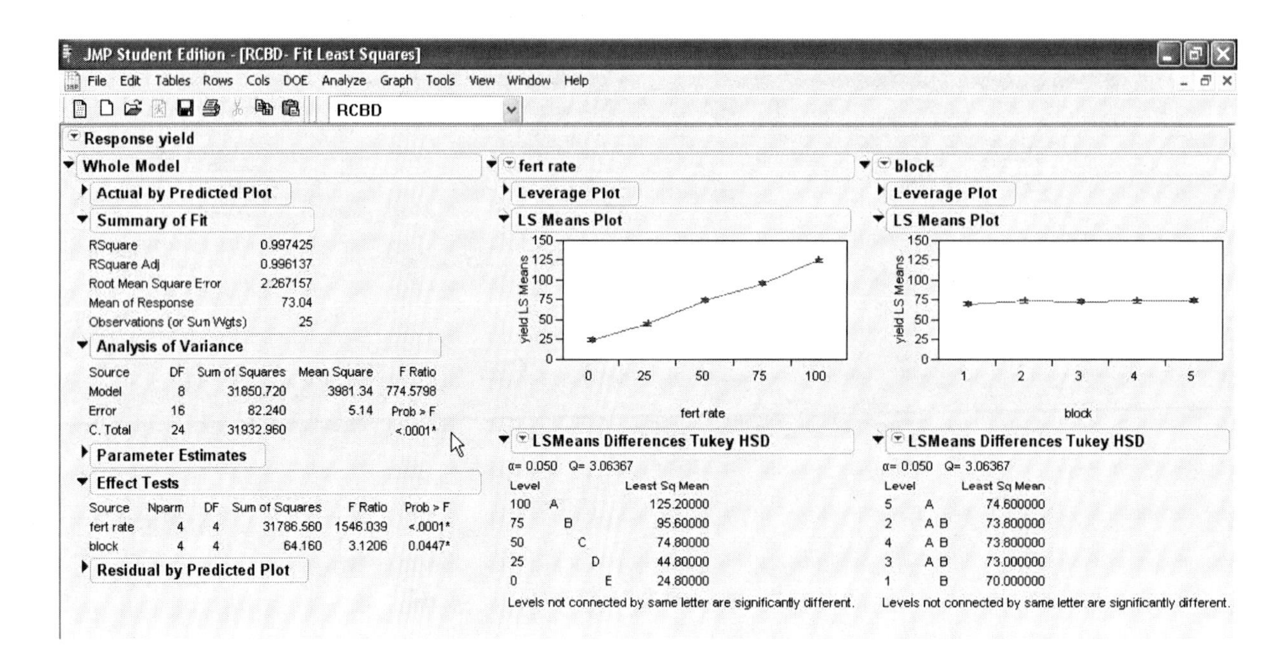

Figure 9-10. Data analyzed with JMP from the RCBD with field blocks of turf grown under various rates of fertilizer.

3. Block #1 should not be avoided in the field because it was not as effective as the other blocks, but the grower should investigate as to why it is not producing as well as the other blocks. Perhaps a soil analysis or drainage study should be implemented.

LS Design Revisited with Dairy Cattle and Milk Yield

Our example of the Latin square design comes from a dairy herd where the producer is trying to determine if a new diet leads to an increase in pounds of milk. Unlike the hog diet example, each cow in the experiment is going to try out all four feeds. A data sheet used to gather LS data for the analysis is shown in Figure 9-11. Each cow had an opportunity to consume each feed during different time periods. If you wanted to analyze these data with JMP, you would open this file with JMP and go to the **Analyze, Fit Model** platform as seen in previous examples.

In this example we see the familiar **Model Specification** box (Figure 9-12) and that the pounds of milk data are continuous and cow, period, and diet labels are nominal. As always, do not forget to click **Run Model.**

The data from Figure 9-13 show there is a significant difference in the overall ANOVA model (p-value of 0.0008), a very high R^2 Adj of 0.921644 (92.2% correlation), and significant differences between all three effects tests: cows, diet, and period. Can you give the p-value for each effect?

Since we probably hypothesized that each cow would eat differently and the time periods would affect the milk yields (perhaps the weather was changing, maybe there was a hot spell), it should not surprise us there were statistical differences between them. We want to know which feed (diet) was associated with the highest milk production (in pounds).

Clearly there was a significant difference between treatments and by looking at the LS Means Plot and Tukey means comparisons in Figure 9-13, we see two groups of diets, diet #3 and #4 were statistically similar (A) and diets #1 and #2 were similar to each other as well (B).

This is a pretty good model, looking at the R^2 adj of 92.2%. This means that 92.2% of our dependent variable milk yield is attributed to the independent variables (cow, period, and diet). If you run the analysis with just milk yield and diet, the R^2 adj sinks down to 16.6% and the overall ANOVA is not significant. Each individual term in the model (cow and time period) really do help to explain our results. This shows that you have to include everything in your analysis in order to do a thorough job.

Would this dairy producer still consider switching diets? Probably so, in this case he or she would most likely choose between diet #3 or #4 since they are statistically the same. The differences between each cow and period were significant (not surprisingly, see LS Means Plots in Figure 9-14). What could be a factor in this experiment is the period fed, as mentioned earlier. Was the weather getting warmer? Was there a freeze? Someone must know or remember something.

Since we know that period and cow play important roles in our model, Figure 9-14 shows the production of pounds of milk by individual cow. Cow #3 had the highest production of the four and cow

HOH Latin Square (LS) Example Data Entry

Cow	Period	Diet	lbs milk
1	1	1	192
2	1	2	195
3	1	3	292
4	1	4	249
1	2	2	190
2	2	4	203
3	2	1	218
4	2	3	210
1	3	3	214
2	3	1	139
3	3	4	245
4	3	2	163
1	4	4	221
2	4	3	152
3	4	2	204
4	4	1	134

Figure 9-11. Data from an Excel file showing a LS design with dairy cattle feed and milk yield in pounds.

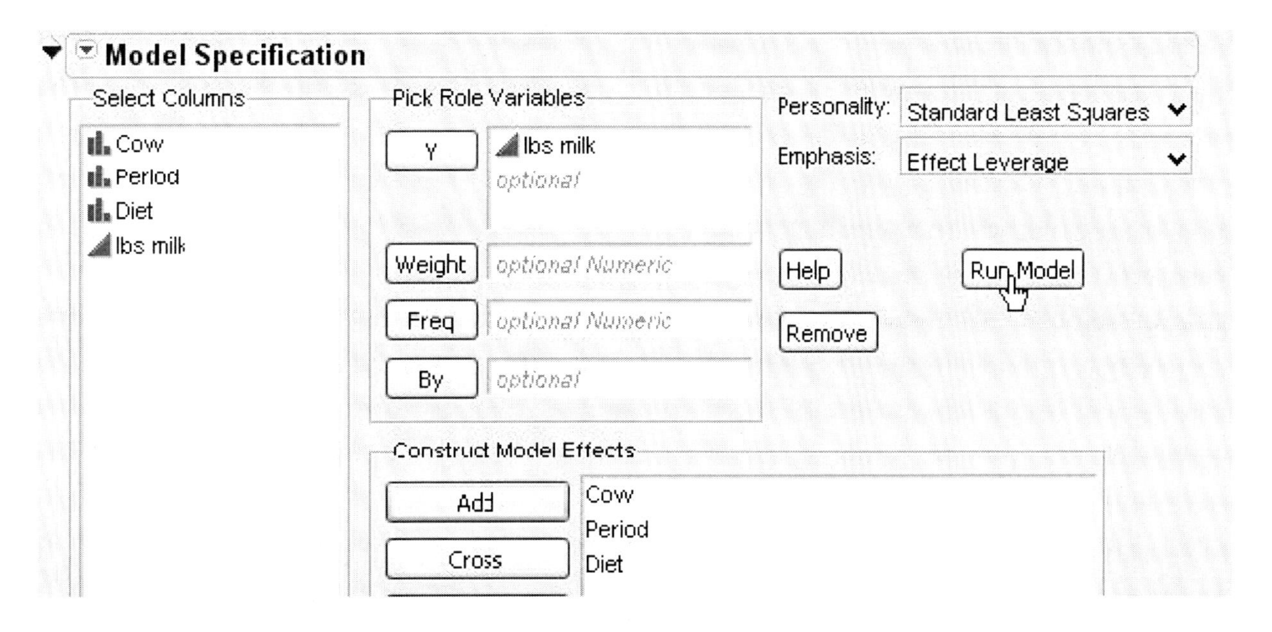

Figure 9-12. The dependent variable (pounds of milk) is being evaluated. Use the Y Role Variable box and the Construct Model Effects boxes in JMP.

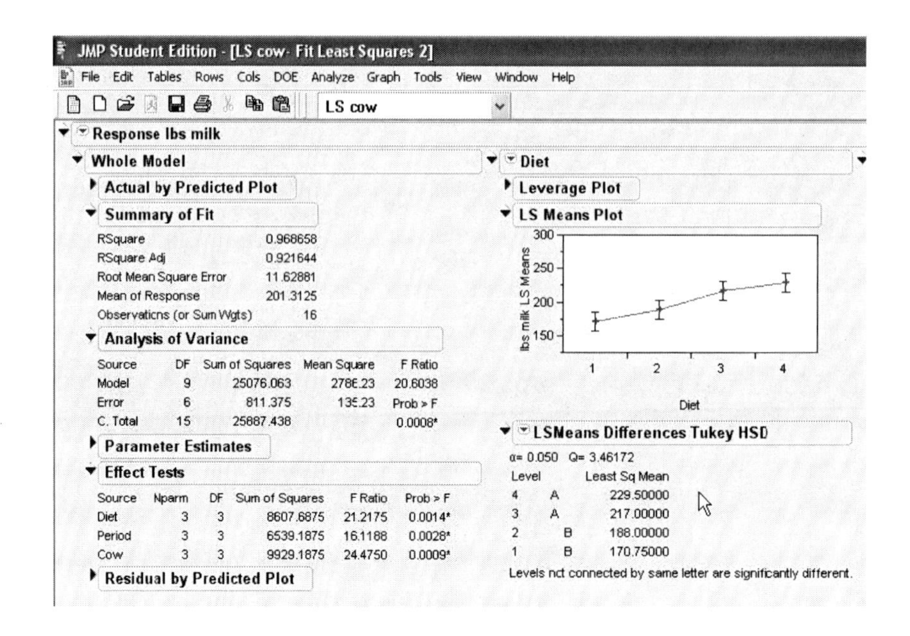

Figure 9-13. Data analyzed with JMP for the LS for diet differences and pounds of milk produced in dairy cows.

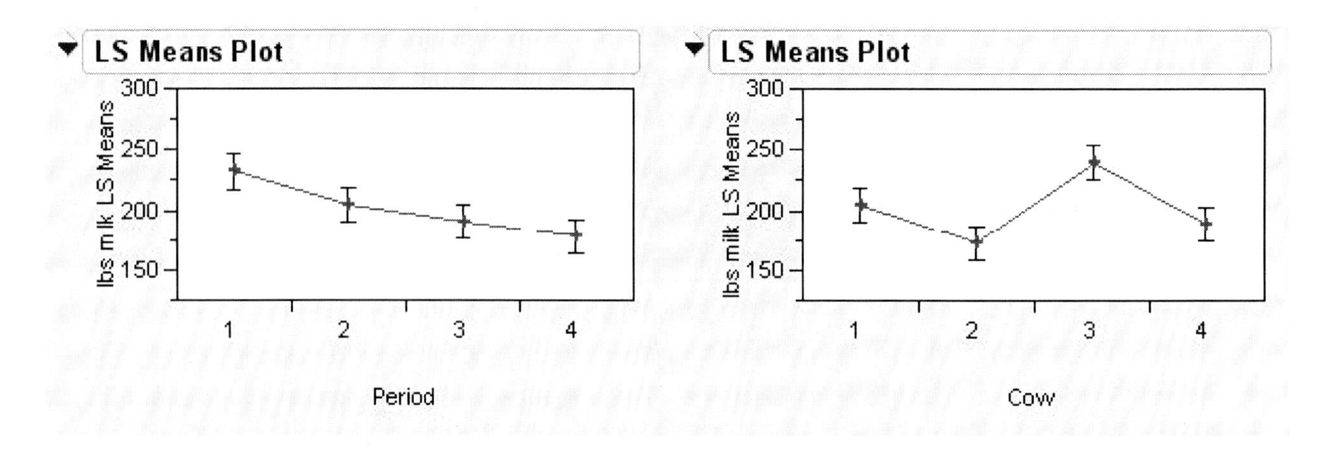

Figure 9-14. The LS Means Plots for period differences and individual cows.

#2 had the lowest. It would be a good idea to see what the pedigrees were on these cattle to ensure that highest-producing cows remained in the herd.

In summary, if you were to discuss these results with the dairy producer involved with this LS analysis, you could include the following in your discussion:

1. The cattle did produce different amounts of milk. If cow #2 continues to have low production, she should be culled from the herd.
2. The whole model was highly associated to milk yield (R^2 Adj of 0.921644, which is a 92.2% correlation).
3. The time period might have had an effect on yield, since milk production went down later on in the trial across the board. What was going on that might have contributed to this decline in milk production? Was it weather-related?

Work on improving your "results and discussion summary" every time you look at data. What recommendations can you make and how will they impact the future of those involved?

Interaction Between 2-Way Factorials

Sometimes we have experiments that involve more than one independent variable and we might wonder if there is a connection between these independent variables. Do they interact; and, if so, is the result enhanced or reduced by their interaction? Interaction between combinations of factor levels can sometime result in unique effects, therefore this arrangement deserves our attention.

A statistical arrangement that is frequently encountered in agriscience research is the factorial arrangement. It is similar to blocking in that there is an additional factor to take into consideration but it's more likely to be seen in CRD.

Factorial arrangements look at different factors or levels of several independent variables and specifically the interaction between them. These different levels or "ways" from two independent variables make up a 2-way factorial arrangement. These look similar to what we have already seen and, in fact, they could have been CRD with the exception of having an additional independent variable (gender, variety, etc.)

Let us look at the 2-way factorial and how there might be interaction between the terms in addition to any regular treatments differences. Figure 9-15 shows a sample data sheet for a really large fertilizer concentration trial between a foliage bedding plant and a flowering one. There are two bedding plant species, five treatments (concentrations), and five replicates/treatment. The dry shoot weight is the dependent variable. Figure 9-16 shows how to analyze interaction effect between independent variables. Again, this is a 2-way factorial because species and treatment concentration are the "ways" or independent variables.

To analyze these data to include the main effects (plant species and fertilizer concentration) select the main effects as seen in previous figures and place them in the Construct Model Effects box (Figure 9-16). Select one of them and click on the button **Cross**, then return to the Select Columns area on the left-hand side and select the other variable you wish to cross or interact with the first variable. In Figure 9-16, **cultivar** was selected first, then **Cross** was selected, then **P fert conc** was selected third. This resulted in the interaction term (**cultivar*P fert conc**) which we will analyze in addition to the main effects in the next step. In this analysis JMP will help us look at the results from P fert conc (phosphorus fertilizer concentration), cultivar (plant variety), and any interaction between fertilizer and variety. Do not forget to click on **Run Model** to tell JMP to perform the ANOVA analysis.

When we begin to analyze these data, we see a strong correlation in the model (R^2 adj is 89.6%). This states that 89.6% of the dry shoot weight can be attributed to the model which includes fertilizer rate, P fertilizer concentration, and interaction between rate and concentration.

In Figure 9-17, we see the ANOVA table which shows an overall significant difference between treatments and the Effects Tests table shows that all three parts of the model (fertilizer concentration, plant variety and the interaction term) were significant. What are each of the p-values from the entire model?

The Tukey table in Figure 9-17 shows there is no significant difference between the fertilizer rates #2–#5 (all earned As) but fertilizer rate #1 (the control) was significantly different than the other four rates with a B, its plants resulted in the lowest in dry shoot weight. If we went no further, which of the rates (between 2–5) would you recommend to a professional greenhouse grower and why? The right answer

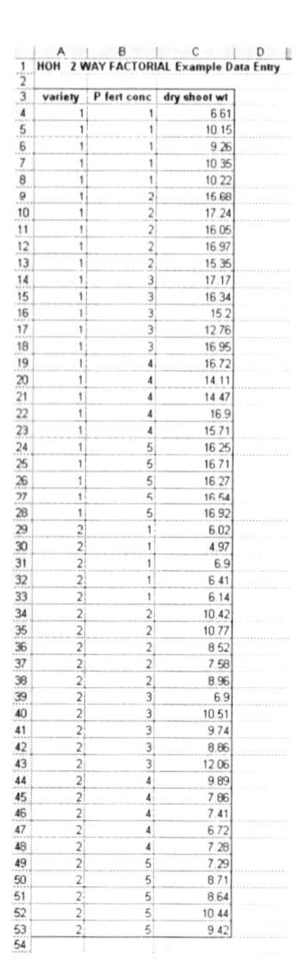

HOH 2 WAY FACTORIAL Example Data Entry

variety	P fert conc	dry shoot wt
1	1	6.61
1	1	10.15
1	1	9.26
1	1	10.35
1	1	10.22
1	2	16.68
1	2	17.24
1	2	16.05
1	2	16.97
1	2	15.35
1	3	17.17
1	3	16.34
1	3	15.2
1	3	12.76
1	3	16.95
1	4	16.72
1	4	14.11
1	4	14.47
1	4	16.9
1	4	15.71
1	5	16.25
1	5	16.71
1	5	16.27
1	5	16.54
1	5	16.92
2	1	6.02
2	1	4.97
2	1	6.9
2	1	6.41
2	1	6.14
2	2	10.42
2	2	10.77
2	2	8.52
2	2	7.58
2	2	8.96
2	3	6.9
2	3	10.51
2	3	9.74
2	3	8.86
2	3	12.06
2	4	9.89
2	4	7.86
2	4	7.41
2	4	6.72
2	4	7.28
2	5	7.29
2	5	8.71
2	5	8.64
2	5	10.44
2	5	9.42

Figure 9-15. The data sheet in Excel is long but keeps you organized.

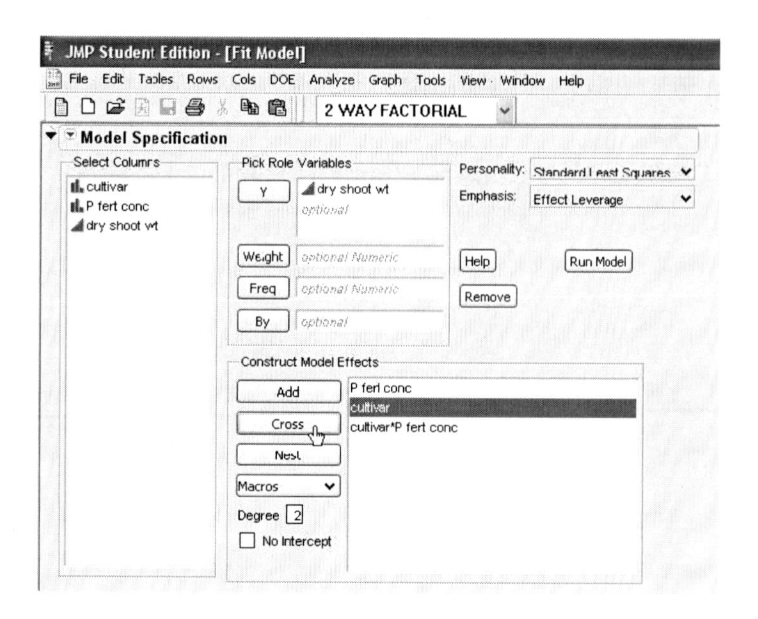

Figure 9-16. There is something new in the Construct Model Effects box, an interaction term was made by crossing cultivar with P fert concentration.

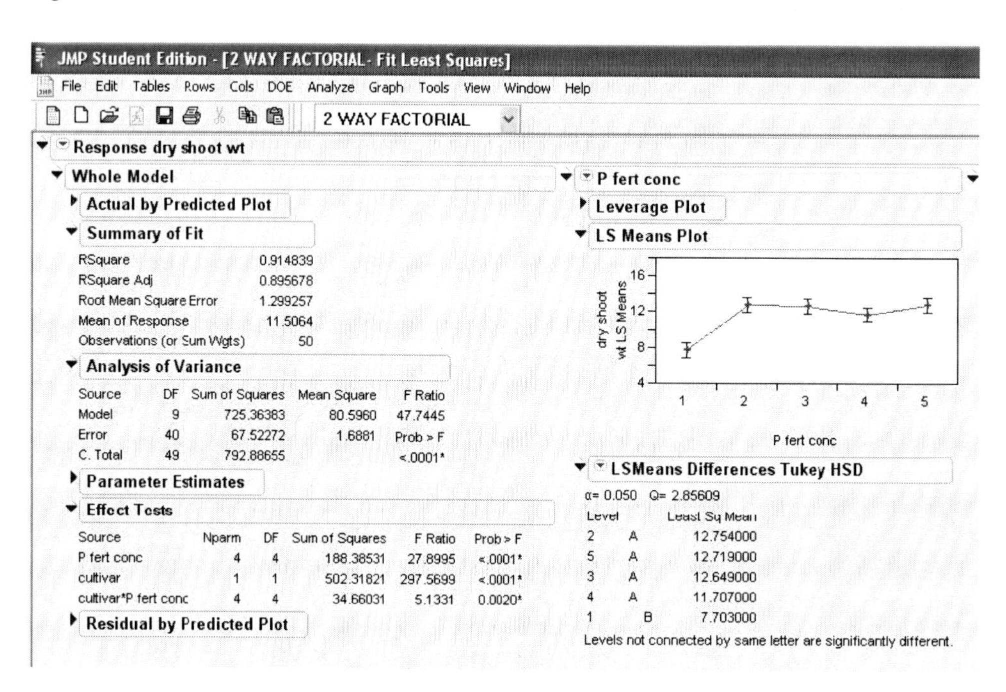

Figure 9-17. The P fertilizer concentration was found to have statistically the same dry shoot weights for 4 out of 5 treatments in both plant varieties.

is the cheapest one since they are all statistically the same. Why pay more money for fertilizer if less will do the same job? It is all about improving your bottom line, your profitability.

Figure 9-18 also has some interesting things to share. The LS Means Plot on the left for plant variety (cultivar) clearly shows that plant cultivar #2 is a much smaller plant, almost half the mass of plant

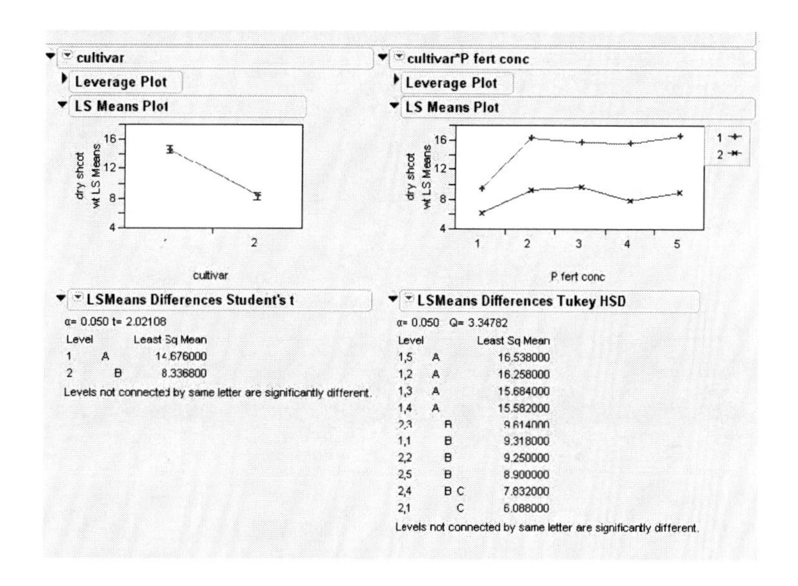

Figure 9-18. There is an obvious difference between the two plant varieties as seen in the LS Means Plots for the interaction term (cultivar*P fert conc).

cultivar #1. We do not know if it is shorter, but since it has a smaller dry shoot weight, there is reduced mass, which could be due to a smaller size. It could be height, but we do not know since we do not have any height data. The interaction term (cultivar*P fert conc) on the right of Figure 9-18 shows some very interesting things.

The interaction was between the independent variables plant variety and P fertilizer concentration. Do different species need or utilize P differently? Should we be fertilizing all plant varieties in the same way with the same amounts of P? Those are good questions, let us look at the data and see what they are trying to tell us.

The LS Means Plot for the interaction term is good for showing trends between two subjects (Figure 9-18), plant varieties in this example. Both varieties grew better with concentrations greater than the first rate and both tend to level off on their growth. More P fertilizer is not providing more growth and, in fact, on both lines we see a dip in dry shoot weight as concentration increases. This suggests that P is not needed by plants in large quantities and that "less is more" when it comes to P fertilizer.

As we look at Tukey for the means comparisons of the interaction, we see that cultivar #1 grows bigger than cultivar #2 (which we already knew) and that both control groups (1,1 and 2,1) were the lowest performers of all tested. Again, the Tukey means comparisons analysis under the interaction term is shown in Figure 9-18 there are no differences between each group of plants (with the exception of the control group) from each cultivar.

This analysis tells us that we can utilize any of the fertilizer rates greater than the lowest concentration when growing plants in these varieties. Smaller amounts of P fertilizer should be tested in the second cultivar to see where the optimum point is located. After that, the next step would be to test additional varieties and see if those data support our hypothesis that P is only used in small amounts in bedding plants.